DESIGNS IN SCIENCE
USING LIGHT

SALLY and ADRIAN MORGAN

Using Light
Copyright © 1993 by Evans Brothers Limited
All rights reserved. No part of this book may be reproduced or utilized in any form or by any means, electronic or mechanical, including photocopying, recording, or by any information storage or retrieval systems, without permission in writing from the publisher. For information contact:
>Facts On File, Inc.
>460 Park Avenue South
>New York NY 10016

Library of Congress Cataloging-in-Publication Data
Morgan, Sally.
  Using light / Sally and Adrian Morgan.
      p. cm. — (Designs in science)
  Includes index.
  Summary: Discusses the physical principles of light, its functions in the natural world, and its integration with modern technology.
      ISBN 0-8160-2980-6
      1. Force and energy — Juvenile literature. 2. Power resources — Juvenile literature. 3. Light — Juvenile literature. 4. Light sources — Juvenile literature.
      [1. Light.]
      I. Morgan, Adrian. II. Title. III. Series: Morgan, Sally. Designs in science.
QC73.4.M68 1993
531—dc20   93-21535

Facts On File books are available at special discounts when purchased in bulk quantities for businesses, associations, institutions or sales promotions. Please call our Special Sales Department in New York at 212/683-2244 or 800/322-8755.

10 9 8 7 6 5 4 3 2 1

This book is printed on acid-free paper.

Editor: Su Swallow
Designer: Neil Sayer
Production: Peter Thompson
Illustrations: Hardlines, Charlbury
              David McAllister

Acknowledgments

For permission to reproduce copyright material the authors and publishers gratefully acknowledge the following:

Cover Ray Ellis, Science Photo Library
Title page Will and Demi McIntyre, Science Photo Library
Contents page Sally Morgan, Ecoscene **page 4** NASA/Science Photo Library **page 6** Phil Jude, Science Photo Library **page 7** Philip Craven, Robert Harding Picture Library **page 8** Donald E Carroll, The Image Bank **page 9** (top) Agence Nature, NHPA (bottom) Adrienne Hart-Davis, Science Photo Library **page 10** Jany Sauvanet, NHPA **page 11** Sally Morgan, Ecoscene **page 12** Elyse Lewin, The Image Bank **page 13** (top) Robert Harding Picture Library (left, inset) Roger Tidman, NHPA (bottom) Sally Morgan, Ecoscene **page 14** (left) Adrienne Hart-Davis, Science Photo Library (right) Hans Reinhard, Bruce Coleman Ltd **page 16** (top and inset) Sally Morgan, Ecoscene (bottom) Dr Morley Read, Science Photo Library **page 17** Sally Morgan, Ecoscene **page 18** Adam Hart-Davis, Science Photo Library **page 19** (left) Malcolm Fielding, Science Photo Library (right) Amy Trustram-Eve, Science Photo Library **page 20** (top) Hank Morgan, Science Photo Library (bottom) Anthony Bannierty, NHPA **page 21** (top) G.I. Bernard, NHPA (bottom) Ken Lucas, Planet Earth Pictures **page 22** Richard Kirby, Oxford Scientific Films **page 23** Sally Morgan, Ecoscene **page 24** (top) A Stewart, The Image Bank (bottom) World View/Bert Blokhuis, Science Photo Library
page 25 Hank Morgan, Science Photo Library **page 26** (top and middle) George Bernard, NHPA (bottom) Sally Morgan, Ecoscene **page 27** (left) Robert Harding Picture Library (bottom) Stephen Dalton, NHPA **page 28** Sally Morgan, Ecoscene **page 29** Sally Morgan, Ecoscene **page 31** (left) G.I. Bernard, Oxford Scientific Films (right) Sally Morgan, Ecoscene **page 32** Sally Morgan, Ecoscene **page 33** Dr Frieder Sauer, Bruce Coleman Ltd **page 34** (left) Sally Morgan, Ecoscene (right) Robert Harding Picture Library **page 35** Sally Morgan, Ecoscene **page 36** (top) Gerald Lacz, NHPA (bottom) John Murray, Bruce Coleman Ltd **page 37** (top and middle) Claude Nuridsany and Marie Perennou, Science Photo Library (bottom) Marcel Isy-Schwart, The Image Bank **page 38** (top) Sally Morgan, Ecoscene (bottom) James Holmes, Science Photo Library **page 39** (top) Jany Sauvanet, NHPA (bottom) McAlpine Helicopters Ltd. **page 40** (left) Robert Harding Picture Library (right) G.I. Bernard, NHPA **page 41** Sheila Terry, Science Photo Library **page 42** US Department of Energy/Science Photo Library **page 43** (top) Kim Steele, The Image Bank (bottom) Sinclair Stammers, Science Photo Library

DESIGNS IN SCIENCE
USING LIGHT

SALLY and ADRIAN MORGAN

Facts On File

NOTE ON MEASUREMENTS:

In this book, we have provided U.S. equivalents for metric measurements when appropriate for readers who are more familiar with these units. However, as most scientific formulas are calculated in metric units, metric units are given first and are used alone in formulas.

Measurement

These abbreviations are used in this book.

METRIC   **U.S. EQUIVALENT**

Units of length

km	kilometer	mi.	mile
m	meter	yd.	yard
cm	centimeter	ft.	foot
mm	millimeter	in. or "	inch

Units of temperature

°C	degrees Celsius	°F	degrees Fahrenheit

Units of speed

km/h	kilometers per hour	mph	miles per hour
km/s	kilometers per second	mi./sec.	miles per second
m/s	meters per second	ft./sec.	feet per second
		in./sec.	inches per second

Units of area

ha	hectare	sq.yd.	square yards
cm^2	centimeter squared	sq. ft.	square feet
mm^2	millimeter squared	sq. in.	square inches

Units of volume

l	liters	cu. ft.	cubic feet
cm^3	centimeters cubed	cu. in.	cubic inches

Using Light is one book in the seven-volume series Designs in Science. The series is designed to develop young people's knowledge and understanding of the basic principles of movement, structures, energy, light, sound, materials, and water, using an integrated science approach. A central theme running through the series is the close link between design in the natural world and design in modern technology.

Contents

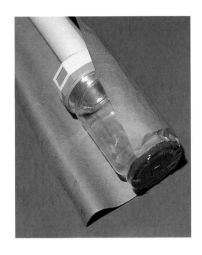

Introduction 6
- What is light?
- The electromagnetic spectrum

Reflecting and refracting light 9
- Mirrors
- Bending light
- Optical fibers

Filters and pigments 14
- Pigments
- Photography and printing

Making light 20
- Electric lights
- Bioluminescence

Capturing light energy 24
- Photochemistry

Light sensors 28

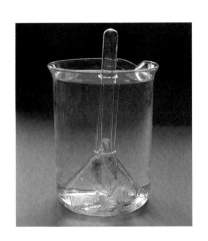

Seeing with light 32
- The eye
- Color vision
- Cameras

Invisible light 38
- Ultraviolet light
- Infrared light

Interference and polarized light 42
- Interference patterns
- Polarized light

The future 44

Glossary 46

Answers to the questions 46

Index 47

DESIGNS IN SCIENCE

Introduction

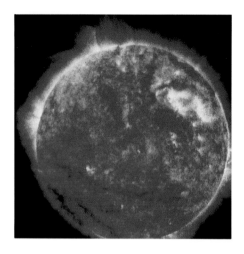

The sun is a ball of gases that burns at an incredibly high temperature, giving off lots of light.

Light brings life to the world. Our main source of light is the sun. It provides us with light and heat and it controls the patterns of life on earth. Plants need light to grow. Animals depend on plants for food, either directly or indirectly. Humans and other animals use light to see by. Without the sun, life as we know it would come to an end.

Light is a form of energy, just like heat and sound. Light energy is frequently given off from hot objects, so we see light when materials such as wood or coal are burned.

Some living creatures, such as fireflies and glowworms, can create light within their own bodies. Humans have learned to create light artificially, using a variety of methods. For example, light can be created by heating metals or by passing an electric current through a gas. But all means of creating light are basically the same: light is given off as a result of atoms within a substance being given energy.

What is light?

Scientists have studied light for hundreds of years, trying to discover what it is made from. In the 17th century Sir Isaac Newton knew that light traveled in straight lines and cast shadows and that it could be reflected off mirrors. He thought that a beam of light was made from a stream of tiny particles. Later, other scientists showed that light could also travel as waves, rather like ripples crossing a pond, except that the waves are tiny electric and magnetic vibrations. Today scientists believe light has a dual

Visible light is just a small part of the electromagnetic spectrum that stretches from long radio waves to very short cosmic rays. The different wavelengths have particular uses.

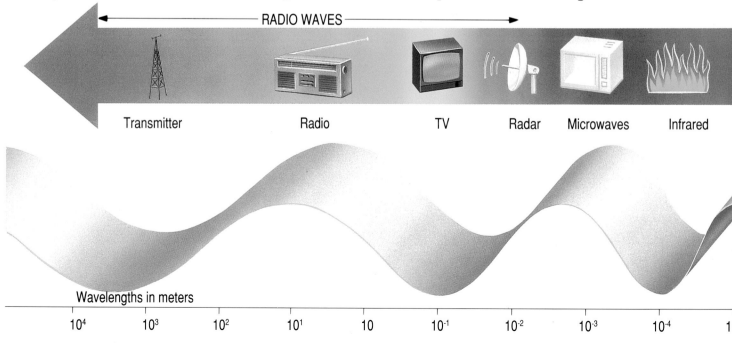

RADIO WAVES

Transmitter — Radio — TV — Radar — Microwaves — Infrared

Wavelengths in meters

10^4 — 10^3 — 10^2 — 10^1 — 10 — 10^{-1} — 10^{-2} — 10^{-3} — 10^{-4}

USING LIGHT

nature, exhibiting some properties of waves and some of "particles," tiny packets of light known as photons.

The waves are not all the same size. The distance between the peaks of two waves is called the wavelength. There are short wavelengths and long wavelengths. When we see colors we are really seeing light of different wavelengths. For example, red light has a longer wavelength than blue light. The amount of energy carried by each type of photon varies too. Photons with a shorter wavelength carry more energy. For instance, blue photons carry more energy than green ones.

The electromagnetic spectrum

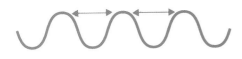

The distance betweeen two neighboring peaks is called the wavelength.

Sunlight is called white light, but it is really a mixture of colors that appear white when mixed together. We can only see a small part of the light given off by the sun, the part known as the visible spectrum. The visible spectrum consists of red, orange, yellow, green, blue, indigo and violet light. There is no definite boundary between one color and the next. Instead there is a continuous change of color from red to violet. You see the visible spectrum when you see a rainbow. As white light passes through droplets of water in the sky, the light is refracted (bent). The longer wavelengths of light are refracted more than shorter wavelengths, which causes the colors to be separated as they pass through the raindrops. Natural sunlight contains a particular mixture of wavelengths of light. Artificial lighting is designed to produce a light that is very similar to sunlight, although the actual composition of wavelengths is different.

There are also wavelengths of light above and below the visible spectrum that humans cannot see. Ultraviolet (UV) light has a slightly shorter wavelength than blue light and infrared (IR) light has a slightly longer wavelength than red light.

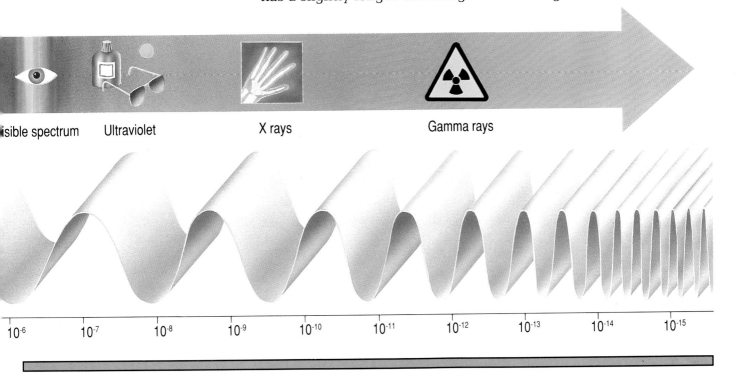

8 | DESIGNS IN SCIENCE

A rainbow is produced when white light is bent as it passes through water droplets.

Light is just one form of wave energy, but there are many others including radio waves and microwaves. These waves are all made from electric and magnetic ripples produced by electrons as they vibrate and lose energy. They are known as electromagnetic waves. All parts of the electromagnetic spectrum travel at the same speed (300,000 km/s or 186,000 mi./sec.), from the longest radio waves to the shortest cosmic rays. The different types of waves can be identified by their wavelengths. The wavelength determines how they will behave, so the waves are grouped according to wavelength. The complete grouping is called the electromagnetic spectrum, of which light is just one small part. Beyond UV light, with even shorter wavelengths, are X rays, gamma rays and cosmic rays. These rays, which have very short wavelengths, carry so much energy that they are very dangerous.

The color of objects comes from the interaction of light falling on the object and the presence of any pigments in the object itself. A pigment is a substance that gives an object a color by absorbing certain wavelengths of light and reflecting others. This is described in greater detail on page 16.

This book studies the ways in which light is used by people, plants and animals. It investigates the nature of light, how it can be trapped and used and how it can be produced.

Important words are explained at the end of each section, under the heading of **Key words** and in the glossary on page 46. You will find some amazing facts in each section, together with some experiments for you to try and some questions for you to think about.

> ! *The longest wavelengths, long wave radio waves, are more than 100km long, but the shortest, gamma rays, are less than 0.000000001 mm long!*

Key words
Photon a tiny unit of light that has a small amount of energy.
Spectrum the rainbow colors produced when light passes through a raindrop or a prism (a specially shaped piece of glass).
Wavelength the distance between two neighboring peaks or troughs of a wave.

USING LIGHT

Reflecting and refracting light

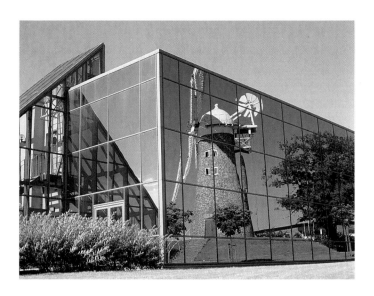

Light normally passes through glass, but at certain angles it is reflected instead.

Light always travels in straight lines, but if a beam of light hits an object in its path it will either pass through it, reflect off it or be absorbed. If the beam of light passes from air to water it may be refracted (bent).

Light can pass through transparent objects without any light being reflected in any direction so it is impossible to see something that is truly transparent. Air, for example, is transparent, as is glass. Sometimes, however, the surface of even a transparent material such as glass can be reflective. At certain angles the light does not pass through the glass, but is reflected instead. It can be very difficult to see what is behind a reflective surface if bright light bounces off it at an angle. For example, if the surface of a shop window is highly polished, it can be difficult to see inside.

However, light cannot pass through most objects. If an object is placed in the path of a beam of light and the light does not pass through it, the object is said to be opaque. The light will be reflected in all directions, from the object's surface. The light is said to be scattered. A shadow will be cast behind the opaque object, where the light cannot reach. Objects are visible because the light reflected from them can be seen by the human eye. You can see the pages of this book because sunlight or electric light is being reflected from the surface of each page.

Translucent objects allow some light through but scatter some light as well. Sunglasses are a form of translucent screen, which controls the amount of light reaching our eyes.

 Translucent things, such as clouds, allow some light to pass through. Can you think of any other examples?

Mirrors

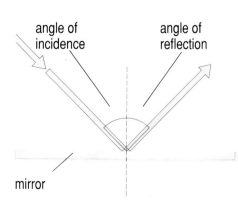

If you look in a mirror, you see an image of yourself. This is because light hits the mirror and is reflected. A mirror is a surface that is smooth enough to reflect a clear image that is an exact copy of the object. This happens because the light rays are reflected in the same direction, not scattered. Your face is the object and your image is reflected by the mirror. A mirror image is inverted; the left side of your face is the right side of the image. Also, the image in the mirror seems to be as far away as the object is in front of the mirror.

Light bounces off a mirror in a similar way to a ball bouncing off the ground. When light hits a surface at a particular angle, it is reflected back at the same angle. We say that the angle of reflection equals the angle of incidence (see diagram).

10 DESIGNS IN SCIENCE

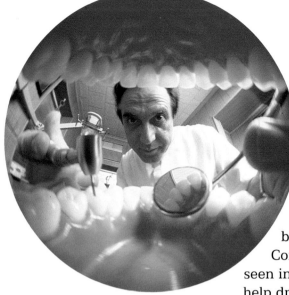

Dentists use concave mirrors so that they can see a patient's teeth more clearly.

> **!** *The largest concave mirror in a telescope is 6m (6.6 ft.) in diameter. It can pick up light from stars millions of kilometers or miles away.*

> **?** *How many types of concave and convex mirrors do you see each day?*

It is not the glass in a mirror that reflects the light, as many people think. In fact, the rays of light from an object pass through a transparent, protective layer of glass before being reflected back by a very thin, shiny metal film on the back surface. A mirror is said to be "silvered," after the metal from which these reflective films were once made.

Most of the mirrors that we use are flat, but mirrors with curved surfaces can be used to make the reflected image appear larger or smaller, depending on whether the curve is concave or convex. If the curve is inward, like a bowl, it is called concave. If it bulges out, it is convex.

Convex mirrors make the image seem smaller, so more can be seen in a convex mirror. Driving mirrors are slightly convex to help drivers to see more of the road behind them. Convex mirrors are also sometimes used as security mirrors in stores, so that large areas can be surveyed at once. Concave mirrors, on the other hand, enlarge the image. They are used by dentists to see teeth and are sold as shaving mirrors. Reflecting telescopes also use large concave mirrors. These mirrors can collect the faint light from very distant stars, enabling astronomers to study the universe more easily.

Mirrors are uncommon in the living world, but one particular fish uses mirrors in a remarkable way. The hatchet fish, which is found only in deep sea water, uses a kind of natural mirror to make itself invisible to predators. Its back is dark, while its sides are covered in large silver, reflective scales that act like mirrors. It also has two rows of light-producing organs on its underside. The fish senses the amount of light falling on its back from above and recreates the same level of light on its own underside. Should any

EXPERIMENT

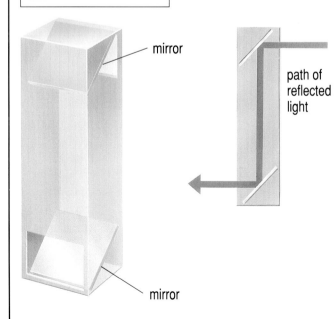

Building a periscope

Periscopes are used on submarines to allow the captain to see ships on the surface. Periscopes also let you see around corners. You can make a simple periscope with mirrors and a tube. You will need two small mirrors, some cardboard or a long tube, pen, tape and scissors. Using these materials, see if you can make your own periscope. The diagram on the left will help you.

Can you answer these questions:
Does it matter how long the tube is?
How could you improve the design of your periscope?
Would painting the inside of the tube black help?

USING LIGHT | 11

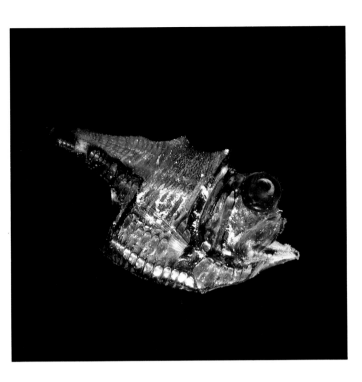

The hatchet fish has mirrored scales along the side of its body as well as light-producing organs on its underside.

predator look up from below the fish, all it will see is an apparently "natural" amount of light. The mirrors on the fish's side reflect back light from the surrounding water, helping the fish to stay invisible. It is a remarkably complex thing to do, for the hatchet fish must, not only reflect the correct amount of light, but also must produce exactly the right color as well. Only blue light can penetrate deep into the sea. The hatchet fish produces the correct color of light by using filters to cut out unwanted colors of light (see page 14).

Nocturnal mammals, such as the bushbaby, need to have eyes that are very sensitive to light since they are active at night when there is very little light. They have large round eyes that resemble bright, silver dishes. These animals have a mirrored layer, called the tapetum, at the back of each eye. This layer reflects light out of the eye again, giving the eye a second chance to absorb the light. The eyes are very sensitive and able to detect even the faintest glimmers of light.

Bending light

These objects appear to be bent due to refraction.

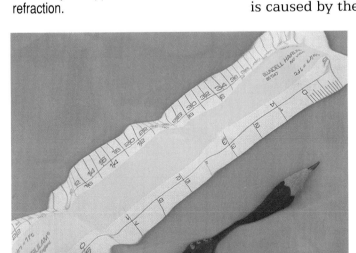

! *Light travels through space at 300,000 km/s (186,000 mi./sec.) but only at 223,000 km/s (140,000 mi./sec.) through water.*

Have you ever noticed how a swimming pool looks shallower than it really is, and your legs look shorter when you are in water? This is caused by the refraction of rays of light. Refraction takes place when light passes at an angle from one transparent medium to another, for example from air to water. If the light hits the water at an angle, it changes direction – it is refracted. The angle is very important. If light hits the water square on, that is to say at right angles, the light will pass straight through and will not be refracted. Light also travels at different speeds in different materials. It can travel much faster through air than water. So when light passes into water its speed is slowed down and the wavelengths get shorter.

You can study refraction by taking a clear glass beaker of water and placing a pencil part in and part out the water. The pencil appears to be bent, even though it is not broken.

Refraction has some very important uses, in particular for lenses in glasses, cameras and microscopes. As the light passes through a lens, its speed changes and the light is bent. By carefully designing the shape of a lens, it is possible to make the

12 DESIGNS IN SCIENCE

The eyes of the four-eyed fish are ideal for seeing above and below water.

light bend by a very precise amount. Opticians, for example, design special lenses to correct sight defects.

Fish have eyes that are adapted to dealing with light underwater. When they are out of the water, their eyes cannot see properly. This is not a problem for most fish, as their eyes never come above the surface. However, the four-eyed fish that lives at the water's surface needs to see both above and below water at the same time! Its eyes are specially adapted to deal with this problem. Each eye is divided into two. The top half is shaped for seeing in air and the lower half for seeing in water. The four-eyed fish swims with the top part of its eyes above the surface and the lower part underwater (see page 32 on the eye).

Optical fibers

! *Optical fibers are less than six thousandths of a millimeter thick, far thinner than a human hair.*

If you have ever looked up while swimming underwater, you may have noticed that you can only see the sky immediately above you. But at a certain angle to the surface, the light is reflected from the undersurface of the water and you only see the reflection of the pool bottom. The light has been unable to escape from the water, instead it has been reflected back into the water. This effect is called total internal reflection, and it occurs because water and air refract light by different amounts. This important effect has been used in designing optical fibers.

Optical fibers are very long, thin fibers of glass. Each fiber is made of two layers of glass. The inner core of glass refracts light more than the outer layer of glass, so any light passing along the inner core of glass cannot escape because total internal reflection occurs at the boundary between the two layers. The light shone into one end of the fiber cannot escape until it reaches the other end. This means that optical fibers are very efficient.

Light travels in straight lines, but the development of flexible optical fibers has enabled light to be transmitted over long distances and apparently around corners. In fact, the light is not actually bent, but it is continually reflected from the boundary of the inner fiber in short straight lines, so that it follows the shape of the fiber.

Optical fibers are ideal for seeing into places that are not easily accessible. Some medical

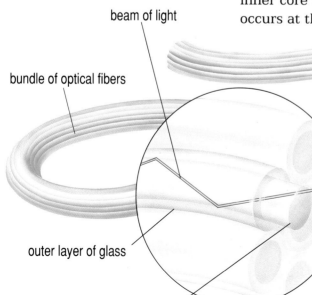

USING LIGHT 13

EXPERIMENT

Trapping light in a jet of water

Total internal reflection can cause light to follow a curved jet of water, just as light passes along an optical fiber. You will need a flashlight, a jam jar with a tight-fitting lid, a 20cm x 20cm (8" x 8") piece of brown paper, a hammer, a nail, a pin, some tape and water. You should do this experiment over a sink.

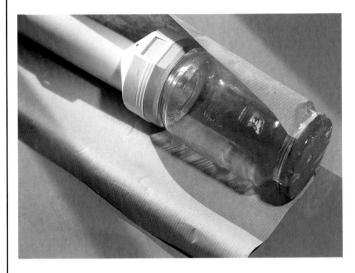

1 Use the hammer and nail to make one large hole (0.5cm or 0.2" in diameter) in the center of the lid of the jar, and make a second smaller hole (0.2cm or 0.16") with the pin near the edge of the lid. Fill the jar with water, then seal the holes with tape.

2 Lay the jar on its side. Wrap the paper around the jar so that it makes a sleeve. Make sure the paper fits closely around the lid, securing it with tape if necessary. Place the flashlight inside the sleeve, behind the jar. Turn the flashlight on so that the light shines into the water. Make sure no light can escape behind the flashlight or at the front near the lid of the jar.

3 Darken the room.

4 Hold the jar (with the small hole at the top) and sleeve in a horizontal position over the sink with one hand and remove the tape from over the holes, using the other hand. The water should escape through the large hole and a beam of light will be trapped within the jet.

 A block of ordinary glass just 1 m (3.281 ft.) thick is opaque, but the glass used to make fiber optics is so pure that you could see through a block 1 km (.625 mi.) thick.

instruments use optical fibers to see inside a patient's body without the need for surgery. For example, an endoscope tube contains fibers so small that they can be passed along a vein to the heart or to some other part of the body. Some of the fibers carry light into the body while others carry the reflected light back so that the doctor can see an image. Instruments similar to endoscopes are sometimes used by engineers to inspect the internal parts of engines without having to dismantle them.

Optical fibers are also used to carry coded light signals over very great distances and so are important to the modern telecommunications industry. An optical fiber cable may contain as many as 27,000 individual glass fibers per mm^2, each one of which transmits many messages. The glass in these fibers must be very pure, as impurities tend to scatter or absorb the light.

Optical fibers are rather like the nerves in our bodies. Just like optical fibers, nerves are very thin and each contains many hundreds of individual nerve fibers bundled together. Nerves do not transmit light but have a similar message-carrying ability. Most of the nerves in our bodies conduct electrical signals from one part of the body to another, usually to or from the brain and spinal cord. It is important that the signal travel quickly and not be altered in any way. Each nerve fiber is designed so that electrical signals cannot escape or change at all, in the same way that optical fibers do not allow light to escape.

Key words
Mirror a smooth polished surface that reflects light in a regular way to form an image.
Refraction the bending of light rays as they pass from one medium to another.
Reflection a wave of light that bounces off an object.

14 DESIGNS IN SCIENCE

Filters and pigments

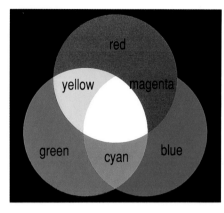

This color wheel shows the three primary colors of light and three secondary colors that are produced by mixing two primary colors.

There are three primary colors of light: red, blue and green. From these, all the other colors can be made. Secondary colors are those produced by mixing together pairs of primary colors. For example, red and green make yellow. There are three secondary colors: yellow, cyan and magenta. White light can be made by mixing all three primary colors together, or by mixing one primary color with the secondary color that is opposite to it in the color wheel. Remember that the primary colors of light are not the same as those of paint (see page 16).

As well as adding colors to produce new ones, colors can also be subtracted from one another. Special filters are used to do this. Filters only allow certain wavelengths of light to pass through. For example, if white light is shone through a red filter, the filter will only let red wavelengths of light through, cutting out all the others. Similarly, a green filter lets through only green light. A cyan filter is a secondary color, so will only let green and blue light through.

There are many examples of natural filters. The ozone layer, high up in the atmosphere, is a filter that blocks out harmful wavelengths of UV light. This layer is very important to life on earth because overexposure to certain wavelengths of UV light, natural or artificial, can cause damage to eyes and skin, which may eventually lead to skin cancers (see page 38).

Our skin also acts as a filter, preventing the UV light from penetrating too deeply. There is a pigment called melanin in the cells of the skin that helps to increase the skin's natural protective ability. If the skin is exposed to the sun, more melanin is produced and this causes the skin to darken and tan.

Some human races have much higher natural levels of melanin and their naturally brown skins give them greater protection from the sun.

In the same way that light skin becomes darker when exposed to UV, special glass has been developed that can do the same thing – only much faster. This photochromic (or photogray) glass is used in some sunglasses. The glass contains millions of tiny silver crystals that form a metallic coating over the glass. In dark conditions the glass is colorless, but when it is exposed to sunlight the crystals become opaque and the glass becomes progressively darker. When the source of light is removed, the silver crystals regenerate and the glass clears. Such sunglasses can change from dark to light in about 10 seconds.

Some tropical tree frogs may look green, but their bodies are actually blue. Blue would show up too much amid the foliage of the plants in which they live, so the frogs have evolved a special outer layer of cells to act as a yellow filter. The yellow filter and the underlying blue skin work together, absorbing all wavelengths of light and allowing just green light to be reflected.

? *Suntan lotion contains a chemical filter. Why do we use suntan lotions?*

USING LIGHT

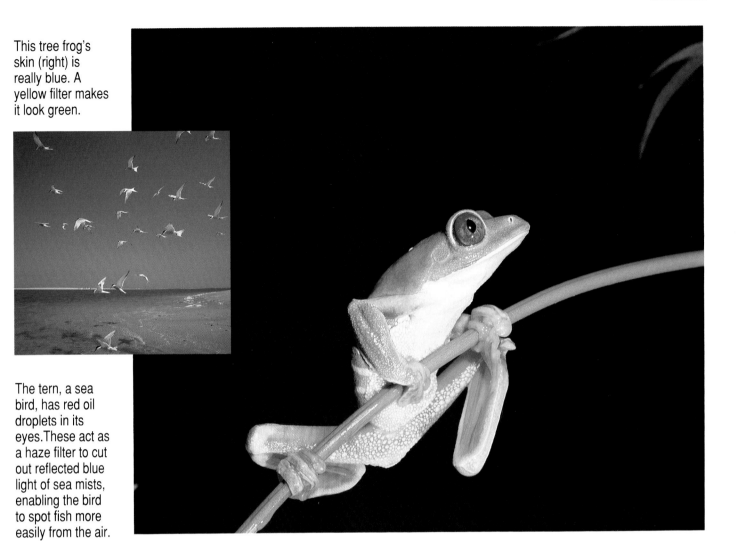

This tree frog's skin (right) is really blue. A yellow filter makes it look green.

The tern, a sea bird, has red oil droplets in its eyes. These act as a haze filter to cut out reflected blue light of sea mists, enabling the bird to spot fish more easily from the air.

EXPERIMENT

A filter box

In this experiment you will test the theory of mixing colors. You will need a shoe box, some tape, scissors, green, blue and red cellophane, a flashlight and a selection of colored objects.

1 Use the scissors to cut out a rectangle from the shoe box lid. Stick a colored piece of cellophane across the opening. You now have a colored window to look through into the box.

2 Cut a hole in the end of the box, just big enough to fit the flashlight.

3 Place a colored object, such as a piece of fruit or a colored playing card, in the box. Replace the lid. Turn the flashlight on and look at the object through the colored window. What can you see?

4 Repeat with other colored objects. What happens to the color of the objects? Which colors change?

5 Now change the color of the cellophane and repeat the experiment using the same objects. Remember that the cellophane acts as a filter. It will allow some wavelengths of light through, but not others.

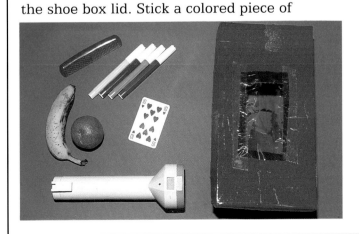

16 DESIGNS IN SCIENCE

Pigments

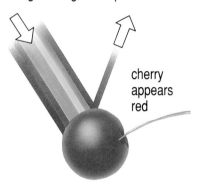

The cherry's red pigment absorbs all wavelengths of light except the red ones.

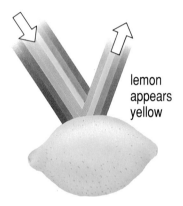

The lemon's yellow pigment absorbs all wavelengths of light except those that make yellow light. Yellow light is reflected from the lemon.

A wide range of natural and human-made dyes are available today.

The pink color of many flamingoes (right) comes from the pigment contained in their food.

Pigments are substances that can give an object color. They are found both in the living world and in human-made materials. Pigments are used in paint to give it color. There are three different primary pigments – blue, red and yellow. Pigments are also used for coloring food, for printing and in thousands of other ways. They produce color by subtraction. For example, a red pigment absorbs all the green and blue light from a white light source, and reflects only red light. New colors can be produced by mixing pigments. Green can be produced by mixing blue and yellow pigments, since green is the only color not absorbed by blue and yellow.

Animals and plants use pigments to give their cells color. Pigments help animals to communicate, to attract a mate and to camouflage themselves.

The cephalopods – the octopus, cuttlefish and squid – have a very elaborate way of communicating using pigment cells in their skin. These cells are filled with different colored pigments, and their size can be controlled by the brain. By making all the red cells large, and all the others small, the animal can produce a red color over its body. This usually indicates anger. The cell sizes can be altered very quickly. Scientists have identified more than 35 different patterns of color used by squid, but they are not sure what all the colors mean.

Many male birds have very colorful feathers during the period of courtship. The color comes not only from pigment within the feathers but also from light reflected by the natural oil that coats the feathers. Some birds look quite drab in wet weather because the water prevents the bright colors from being reflected from the surface of the feathers.

Many animals' pigments act as warning colors to other animals. The vibrant colors of the poison arrow frogs of the rain forest advertise the fact that the animal is very poisonous. Warning colors are found in other organisms too. The poisonous fly agaric toadstool, for example, has a red cap with white spots.

The earliest pigments used by people were those that occurred naturally. Some were found in the earth and others could be easily extracted from plants or animals. Walnut skins provide a strong brown stain while cochineal, a bright red, comes from the body of an insect. Although many natural pigments are still in use, synthetic pigments are now widely available. Aniline, a chemical extracted from oil and coal, is the basic component of many bright synthetic pigments.

USING LIGHT

Photography and printing

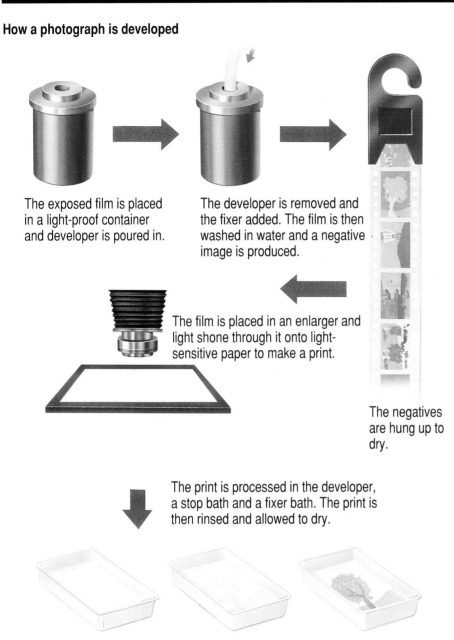

How a photograph is developed

The exposed film is placed in a light-proof container and developer is poured in.

The developer is removed and the fixer added. The film is then washed in water and a negative image is produced.

The film is placed in an enlarger and light shone through it onto light-sensitive paper to make a print.

The negatives are hung up to dry.

The print is processed in the developer, a stop bath and a fixer bath. The print is then rinsed and allowed to dry.

Many of the ways in which people communicate involve the use of light and color, such as printing and photography. Photographic films and papers make use of a group of chemicals called silver halides, which are very sensitive to light. Silver bromide is the most commonly used, mainly for making photographic film. Tiny grains of silver bromide are spread evenly through a transparent liquid to form an emulsion. This is spread thinly over a transparent strip of plastic and allowed to dry. When a photon of light hits a molecule of silver halide, it converts a little of the silver halide into silver metal. Then the film is developed and treated with a chemical solution to "fix" the film, so that it is no longer sensitive to light.

The result of this process is a negative. When you look at a negative, the part of the image that was exposed to the most light, and so created the largest number of silver atoms, appears as a dark area, while unexposed areas are white. In other words, the areas that were dark in the scene come out as light areas on the film, while the lightest areas come out dark. Between these two extremes are many shades of gray where the grains have received intermediate amounts of light.

The human eye and photographic film do not respond to the wavelengths of light in the same way. In photography it is essential to balance the differences between the eye and the film. The human eye can only see the visible spectrum, but modern film emulsions are also sensitive to UV (ultraviolet) light. This can cause problems, especially when photographing out-of-doors. Scenes containing water or snow can often cause a blue haze to appear on color film, unless a UV light filter is used (see page 38). The eye and the film also respond differently to individual colors. The human eye is particularly sensitive to green and yellow but is

18 DESIGNS IN SCIENCE

To produce a black-and-white print, a beam of light is shone through the negative onto light-sensitive paper. The dark areas on the negative (inset) will appear light in the photograph.

less sensitive to deep blue or red. However, silver halides in photographic film react strongly to blue and weakly to red. To produce a black-and-white film that is equally sensitive to the whole visible spectrum, synthetic dye molecules must be attached to the silver halide grains. The dye molecules absorb different amounts of light energy from the visible spectrum and transfer it evenly to the halide.

Color film is made in much the same way as black-and-white film except that three separate layers of emulsion are used. One layer is sensitive to red, one to blue and one to green.

Color printing mixes a limited number of colors together in a way that produces a wide range of colors. The most common printing process uses three basic colors – cyan, magenta and yellow – as well as black, to produce all the colors of the visible spectrum. All the colors reproduced in this book were made from these colors. The original image has to be separated into three images, one in cyan, one in magenta and one in yellow. A fourth image will also be made in black. Black is used for the text as well as pictures. The separations are made by photographing the image through three filters of the primary colors of light (red, green and blue). For example, when a blue

Three separate images in cyan, magenta and yellow (left) are combined with a fourth image in black to form one full color image (above right).

USING LIGHT 19

IR-sensitive films are made by attaching a dye molecule to the silver halide. The photograph on the right was taken with normal film. The one below was taken using IR-sensitive film. How many differences can you spot?

 Why is it safe to develop a black-and-white film in red light, but not in blue?

Key words
Filter a transparent substance that will only let through light of one or more particular wavelengths.
Pigment a substance that gives a material a particular color, for example a red pigment absorbs all wavelengths of light except red, which is reflected.

filter is used, all the red and green light (which together make yellow) is absorbed and only blue light passes through on to the film. When this is processed, a negative is produced. A print is made of this negative and the only color that will be seen on the print is yellow. The same process is repeated using a red filter to produce the magenta separation and a green filter for the cyan separation. The black separation is made using a yellow filter. The next stage involves "screening" the separations. A glass screen covered in a pattern of tiny dots is placed between the camera and the film. This breaks down the image into thousands of dots. These dots are then engraved on a printing plate. The dots from each separation are carefully superimposed upon each other in the printing press to make a full color reproduction of the original image. It is possible to see the tiny dots of color that make up a colored picture by using a strong magnifying glass.

It is very difficult to reproduce exactly the colors of the original because the inks are not identical matches to the original colors. Some colors are very hard to reproduce, especially pale colors such as very light pink. The colors often have to be adjusted to get the balance right, by adding a little more magenta, for example.

Nowadays electronic scanners are replacing photographic separations. The scanner scans the colors that make up the original image and reproduces them electronically. This method is much quicker and adjustments to the balance of color are easier to make.

Making light

Everything that produces light does so because of the action of atoms. It does not matter whether the light comes from the sun, a bonfire, an electric light or a firefly. They all produce their illumination in the same way.

An atom consists of a heavy nucleus containing neutrons and protons, surrounded by a cloud of electrons orbiting around the nucleus. Each electron can only exist at certain specific levels around the nucleus, so it has a particular place in the electron cloud. If an electron is given extra energy, it will absorb the energy, become excited and "jump" to a higher level. This energy can be provided chemically, by electricity or by some other means. The electron is very unstable in this new, higher position. After a short time, it will return to its original, lower and stable position. As it does so, it gives off light energy in the form of photons. Electrons in different atoms give off light of different wavelengths. By carefully choosing the substance that will be used to produce light, different colors of light can be created.

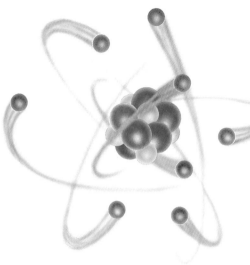

The oxygen atom is made up of a nucleus with eight protons and eight neutrons surrounded by a cloud containing eight electrons.

Electric lights

At night our cities are lit up by thousands of electric lights of all colors. These lights can even be seen from space.

Electric lights are used every day as an artificial form of light. There are several different types. Most common are tungsten filament light bulbs, tungsten halogen bulbs and fluorescent tubes.

The tungsten filament bulb contains a thin metal filament. When electricity flows through the filament, it gets hot because of the resistance of the wire to the passage of the electricity and glows yellow-white. Tungsten is an ideal metal for the filament because its melting point is 3410 °C (6170°F), the highest melting point of any metal. The wire is often coiled so that a greater length will fit into the bulb, thereby giving off more light than a straight filament. When tungsten gets hot in air, it reacts with oxygen; it is oxidized and slowly evaporates. To avoid this, the tungsten is enclosed by a glass bulb and surrounded by a gas such as argon or nitrogen. These gases do not react as much with the tungsten, so the life of the bulb is

USING LIGHT 21

Tungsten lights are very common in the home as they are cheap and easy to use. The halogen bulb on the left of the picture has a tungsten filament surrounded by iodine gas.

Colorful neon lights are often used in advertising displays.

Which is the most common type of light in your home?

The filament of a tungsten bulb reaches temperatures of 2500°C (4532°F).

There are a number of energy-efficient light bulbs on sale today. What types of lights are they for and what are their advantages?

prolonged. However, the process of oxidation is not stopped, it is only slowed down. As the tungsten evaporates, the wire gets thinner and eventually the filament breaks. Tungsten filament bulbs have a life of about 1,000 hours. They are cheap to make and very easy to use, because they can be connected directly to the electricity supply. However, the efficiency of turning electricity into light using such bulbs is only three percent; the rest of the energy is lost as heat.

By surrounding the filament with bromine or iodine gases, known as halogens, the oxidation of tungsten can be further reduced. This means the filament can be heated to a higher temperature and this results in a brighter light. Because halogen bulbs get so much hotter than conventional tungsten bulbs, they would melt ordinary glass, so quartz is used instead. Tungsten halogen bulbs have a 2,000-hour life, but they are still only five percent efficient.

More energy-efficient lighting makes use of a property of the gas itself to produce light. Some gases give out light when high voltage electricity is passed through them while they are at a low pressure. The color of the light given off depends on the gas. Inert gases, such as argon and neon, are most often used and are quite safe. All these lights are called "neon" lights, but neon itself is only responsible for the bright red. These lights are commonly used for advertising. They consist of only a single color per tube, but multicolored neons will soon be widely available. By varying the electric current, up to five colors can be created in a single tube.

Fluorescent tubes are widely used in the home and office. But the light you see from them does not come directly from glowing gas or vapor. When an electric current is passed through mercury vapor inside a glass tube, it gives off UV light. This light is absorbed by phosphor powder, coated on the inside of the tube. The phosphor fluoresces (see page 39), and so it is the phosphor

22 DESIGNS IN SCIENCE

Sodium lights have a distinctive yellow glow. This picture shows sodium lighting in a special growth chamber for plants.

that gives out the visible blue-white light. To get a softer yellow-white light, other chemicals can be added to the phosphor powder. These lights have a life of 5,000 hours and an efficiency of about 25 percent, but they cannot simply be plugged into the electricity supply. They need special electronic starters to produce the high voltage needed to start the light. The starter produces the characteristic flicker seen when a fluorescent light is switched on. Alternatively, metal vapors can be used instead of a gas. For example, mercury vapor produces a blue light, while sodium vapor glows yellow. Sodium vapor lights are sometimes used in street lighting. Their characteristic yellow glow provides a good light in foggy conditions.

Bioluminescence

Bioluminescence means the ability of living organisms to emit (give out) light. All organisms, including ourselves, emit IR light, but some animals, such as glowworms and fireflies, can also emit visible light. They do not use electricity to produce light, but rely on chemical reactions within the cells of their bodies.

This ability to produce light is most common among insects that fly at night, as they use the light to communicate. Both glowworms and fireflies have organs that produce a chemical called luciferin, which reacts with oxygen to produce an eerie glow. The female European glowworm is wingless, with a huge light-producing organ on the underside of her abdomen. When she wants to attract a mate, she curls her abdomen up over her head to reveal the light. It acts as a beacon to the male.

The female glowworm uses her light-producing organ to attract a mate.

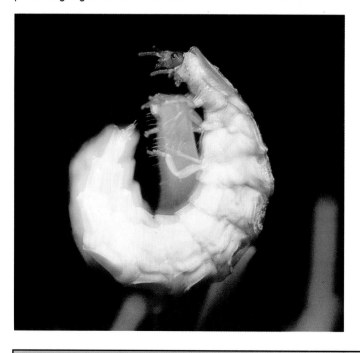

American fireflies, which are similar to European glowworms, behave in a different way. In this case the female can fly, but at night she sits on a branch watching for passing males. As they fly around, the male fireflies switch their lights on and off. When the female sees the light from a male, she responds by flashing her own signals. Each species has its own pattern of flashes, so individuals can signal to the right partner.

The most spectacular displays of light are produced by fireflies in the Malaysian mangrove swamps. Again the displays are designed to attract a mate, but here only the males can produce a light. The fireflies gather together on certain bushes and trees and synchronize their

USING LIGHT 23

Some insects, such as the giant click beetle, can produce two colors of light, one on its head and one in its abdomen.

> **!** *The light of the giant click beetle is so bright that when British troops first landed on Cuba they thought that the great number of lights moving in the woods were the enemy Spanish troops they were fighting. So they retreated and moved on to Jamaica!*

 Not all night-flying insects use light to communicate. How do insects such as moths communicate?

Key words
Atom the smallest part of a chemical element that can exist and still retain all of the element's properties.
Bioluminescence the giving off of light by living organisms.

flashes so that they all flash at the same instant. The whole bush is lit up for a second. This powerful light source can be seen for hundreds of feet and is easy for the female to see, whereas the light from a single male would be lost in the thick vegetation.

The light-producing organ of the firefly is quite complex. It is composed of three layers. The outer layer is a transparent part of the abdominal wall. Inside this is the light organ itself, a series of elongated light cells. The inner layer is formed from opaque cells filled with granules of uric acid. It acts as a reflector (see page 9 on mirrors). Nerves control the amount of light that is produced. The color of the light varies between species but is most commonly yellow. In contrast the giant click beetle, another species of insect, is multicolored, having two green lights behind its head and an orange light in its abdomen.

In the very deepest parts of the ocean there is no natural light, so many deep sea fish produce their own light. The angler fish and the flashlight fish have special light-producing bacteria in their bodies. The angler fish has a spine with a bulb on the end that it dangles in front of its mouth. The bacteria in the bulb only glow when they are supplied with blood rich in oxygen. By increasing or reducing the flow of blood to the bacteria, the fish's light can be switched on or off. The bulb acts as a lure, attracting smaller fish to the angler fish's huge mouth.

Light produced by animals is made chemically. It is a cool light and is more efficient than the electric light, as the process does not generate any waste heat. Scientists have been able to reproduce these chemical reactions, and there are now lights available that work in just the same way. They contain chemicals that, when shaken together, emit a green light, rather like that of the glowworm. They are very useful for emergency situations – and can be fun for parties. But once the chemicals are all used up, the light is extinguished and cannot be renewed.

Flashlight fish use their lights to keep the school together as they swim through the dark seas. They can quickly turn out the lights if a predator approaches.

Capturing light energy

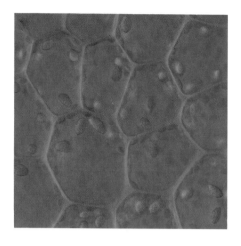

Photosynthetic cells in a leaf contain tiny round, green discs called chloroplasts, which are packed with chlorophyll.

The sun is like a massive nuclear power station that radiates energy into space. Only one thousandth of one millionth part of the sun's energy actually reaches the earth as light and heat. About 30 percent of this energy is reflected by the atmosphere straight back into space. The rest is absorbed by the atmosphere, land and oceans.

A small amount of the sun's energy that falls on the earth is trapped by green plants. Plants use light energy to make their own food. Without plants, no animals could survive, for plants are always at the bottom of the food chain. Herbivorous animals, such as antelopes, cows and snails, obtain the energy they need from the plants on which they feed. Herbivores are directly dependent on plants. These animals in turn are fed on by larger carnivorous animals such as lions, foxes and predatory birds.

To use the energy from the sun, plants must be able to trap light and convert it into a form of energy capable of being transported within the plant or stored for future use. Plants are energy transformers, for they can convert light energy into chemical energy. This chemical energy may then be converted into yet another form of energy.

The process of converting light energy into food is called photosynthesis. The word means "making something using light." It is quite a complex process, but it can be summarized as:

light + carbon dioxide + water → food (glucose) + oxygen

Plant leaves act like solar panels, trapping the light that falls on the plant. The leaves are green because they contain a green pigment called chlorophyll. Chlorophyll is found in tiny green discs called chloroplasts, which are present in many cells within the leaf. It is chlorophyll that enables the plant to photosynthesize. The plant also needs water, which is absorbed by the roots and taken up the stem to the leaves, and carbon dioxide that diffuses (moves) into the leaf from the air. When light falls on a molecule of chlorophyll, the chlorophyll absorbs the light energy and uses it to convert carbon dioxide gas and water into glucose and oxygen. The glucose is then used by the plant as food or changed into starch for storage within the plant, while the oxygen enters the air.

Oxygen is an important by-product of photosynthesis. Oxygen is required by most living organisms for cell respiration – a process that releases the energy locked up in food. Respiration uses oxygen and produces carbon dioxide.

! *The oxygen produced by the very first green plants killed many of the organisms living on earth at the time. If these organisms could not adapt, they became extinct.*

The leaf is the site of photosynthesis.

sunlight

water from roots

carbon dioxide from air

USING LIGHT 25

 Almost all the oxygen in the earth's atmosphere has been produced by green plants photosynthesizing.

Humans respire and produce carbon dioxide, which we breathe out. Plants also respire. Like animals they respire 24 hours a day. During daylight some of the oxygen made in photosynthesis is used by respiring cells. The rest leaves the plant and goes into the surrounding air. The carbon dioxide produced in respiration is used by the plant in photosynthesis. At night plants do not photosynthesize. Now the oxygen is taken from outside the plant. The carbon dioxide produced in respiration leaves the plant. So, at night a plant is a net producer of carbon dioxide, but during the day it is a net producer of oxygen.

Leaves are specially designed to make sure that photosynthesis takes place as quickly and efficiently as possible. They are arranged to catch the maximum amount of light and they do not shade each other too much. If you stand underneath a

EXPERIMENT

How fast do plants photosynthesize?

When plants photosynthesize, they produce oxygen. By measuring the amounts of oxygen being produced, it is possible to find out how fast the plants are photosynthesizing. Pondweeds produce bubbles of oxygen that are easy to see and collect.

You will need some pondweed, a wide-necked jam jar or beaker, a plastic funnel, a small test tube and a ruler.

1 Fill a sink with water and place the jam jar upright on the bottom of the sink.
2 Take two 10 cm (4") sprigs of pondweed. Trim 1 cm, about 1/2," off the end of each stem and push the sprigs into the jam jar.
3 Hold the pondweed in position as you place the funnel in the jam jar. Put the test tube under the water. Make sure it is full of water with no air bubbles. Carefully move it under the water so that it is placed over the end of the funnel.
4 Remove the plug from the sink. Hold the jam jar and test tube in place as the water drains away. You should be left with a jam jar and test tube full of water.
5 Carefully lift everything out of the sink and place it on a bright window ledge. When the pondweed is in bright light, it will start to photosynthesize and bubbles will be seen. These will collect in the test tube. You can measure the rate of photosynthesis either by counting the number of bubbles you see rising up the test tube in a minute or by timing how long it takes to produce 1 cm, about 1/2," of oxygen gas in the tube (you can measure this on a ruler).
6 Now repeat this experiment again, but this time place the jar in a darker place. How long does it take the pondweed to produce 1 cm or 1/2" of oxygen now?

You could also investigate the effect of red and blue light on the rate of photosynthesis. You could cover the apparatus with red or blue cellophane or position it in a dark place and use red and blue light bulbs to provide the light (see page 16).

The leaves of a tree are carefully arranged to catch as much sunlight as possible. Very few leaves completely overlap one another.

tree and look up through the branches, you will see that very few leaves overlap. Their large surface area makes sure that as much light as possible falls on the surface of the leaf. Most leaves are very thin, so carbon dioxide does not have to travel far to reach the photosynthesizing cells. Many leaves have a stem called a petiole to enable the leaf to change its position and follow the sun's rays throughout the day.

People use solar panels or collectors to trap sunlight. In many ways, they are similar in design to plant leaves. Solar panels have a large surface area so that they can catch as much light as possible. The panels are positioned so that they do not overlap each other and are pointed toward the sun so that sunlight falls on them for as long as possible. In the northern hemisphere, this means that solar panels are placed on south-facing walls and roofs.

If a great many solar panels are connected together, sufficient heat can be trapped to generate electricity on a commercial scale. In one design of solar power plant, the sunlight falls on concentric rings of mirrors that reflect the sunlight to a central boiler where water is heated and turned into steam. The steam then powers turbines, used to generate electricity.

Solar panels are designed to absorb the heat energy from sunlight. But there are other designs, called solar cells, that absorb the light energy of sunlight and generate electricity directly. They are sometimes referred to as photovoltaic cells. Solar cells are the simplest and least polluting way of generating electricity. Solar cells make use of wafers of silicon that can transform light directly into electricity. The brighter the light, the more electricity produced. Many modern pocket calculators contain solar cells to provide the tiny amounts of electric power needed to make them work. Light shining on the solar cell causes an electric current to flow. These calculators will continue to work for as long as the calculator is held under a light source.

Solar power is very useful in remote places. This "lighthouse" has four panels made up of photovoltaic cells.

USING LIGHT | 27

The world's largest solar power station is in the Mojave Desert in California. The energy from this power station could be used to power 2,000 homes.

Photochemistry

Recent advances in the field of photochemistry have shown how light energy itself, and not just the heat energy that usually accompanies it, can be used to excite atoms in many different substances. One potential use for this is in the production of hydrogen gas.

Hydrogen could be a very useful, pollution-free fuel since it only produces water when it is burned. Although hydrogen is one of the most plentiful substances on the earth, being a constituent of water (H_2O), there are no natural supplies of hydrogen gas. It can be made by passing an electric current through water. This is called electrolysis. The electric current causes the water to break down into oxygen and hydrogen gases. But this process uses a lot of electricity and is very expensive. However, water will also break down when it is exposed to UV light (see page 39). This is called photolysis. This reaction occurs naturally in photosynthesis. In photosynthesis the light causes hydrogen atoms to be formed from water. They combine with carbon from carbon dioxide to form sugars. Photolysis happens quickly inside plants, but elsewhere it is a very slow reaction. Scientists are hoping to use photolysis to make large quantities of hydrogen. One way of making the reaction occur more rapidly is to use a catalyst. A catalyst will increase the speed of the reaction without actually taking part itself. It is unchanged by the reaction and can be used over and over again. Titanium oxide, mixed with tiny amounts of platinum and sodium carbonate (baking soda) has proved to be an excellent catalyst. Hydrogen can now be made by placing water containing the catalyst under UV light.

! *With the right mix of chemicals, 100 liters (260 gal.) of liquid hydrogen could be produced photochemically on a sunny day – enough to run a small car for a few days. The more sun, the more hydrogen can be produced.*

Key words
Photochemistry a process that uses light to cause a chemical reaction.
Photosynthesis the process by which plants make food using light energy.

Light sensors

Light affects living organisms in many different ways. Many animals and plants need to be able to detect the length of daylight and hence the time of year in order to breed, produce flowers or migrate at the right time. The day length also affects people. Humans are sensitive to the daylight. We can also construct light meters that are able to measure the amount of light. This allows photographers, or the camera itself, to set the correct exposure for a photograph (see page 36).

Light is essential for photosynthesis. Plant shoots will naturally grow toward a light source. This response to light is called phototropism. The amount of light that a plant receives can also affect its flowering. The timing of flower production is very important, especially for plants that rely on insects for pollination. If the plant opens its flowers at the wrong time of year, it may not be able to produce seeds. Most plants in temperate regions of the world, such as most of North America, flower in summer, when there are long hours of daylight. However, others, like the Christmas cactus and the chrysanthemum, only flower when the days are short and the period of darkness long. These plants come from different parts of the world and rely on different insects to pollinate them.

Some flowers such as the crocus open their flowers in the morning (above) and close them in the evening.

Plants can react to day length because they have a special light-sensitive chemical in their leaves, called phytochrome. The levels of this chemical in cells can control many functions of the plant such as flowering.

The prayer plant's leaves are horizontal during the day and vertical at night.

The opening and closing of some flowers' petals is also controlled by sunlight. In the morning, when sunlight starts to fall on their petals, the flower opens. In the evening, when the sun sets, the flower closes. In this way, the delicate reproductive organs of the flower are protected from damage or the frost that can be caused by low night-time temperatures. This form of response to sunlight is called photonasty.

USING LIGHT 29

Arctic summers are very short, although the days are very long. In midsummer, the sun never falls below the horizon and there are 24 hours of daylight. One plant, the arctic avens, has a flower that is exceptionally well adapted to these conditions. The flower is shaped like a parabola. This special shape causes any parallel rays of light hitting the petals to be reflected to the same point of focus (see page 27). The sunlight is focused at the middle of the flower where the reproductive organs are found, and the concentrated light raises the air temperature within the flower by more than 10°C (18°F). This extra warmth attracts the insects that pollinate the flowers. To maintain this temperature, the flowers turn to follow the sun as it moves across the sky, like a radar dish tracking a satellite. These miniature tracking stations are never more than 2° off course.

Some animals can respond to light in their surroundings. The chameleon, a reptile, is an expert at camouflage. It has the ability to change its skin colors to match those of its background. Although the main stimulus for the color changes is light, it is believed that heat and other stimuli are probably also involved in the process. The chameleon has four layers of pigment cells, each containing a different pigment: black, blue, white and yellow. In order to change color, the brain sends messages to the pigment cells in the skin. The pigment cells change their size and thereby alter the colors and patterns of the chameleon's skin.

Human-made materials are not quite so versatile as the skin of the chameleon, but a new type of glass, covered by a thin

The flower head of the sunflower tracks the sun as it moves during the day.

The chameleon has excellent camouflage skills, changing the color of its skin to blend with its background.

The chameleon is not the only animal able to change color. Do you know any others? Why do they change color?

30 DESIGNS IN SCIENCE

Thermochromic pigments could be very useful. How many uses can you think of for them?

coating of a special material, can be turned blue by passing electricity through it. This glass is called electrochromic. It has many novel uses, from car sideview mirrors that can be adjusted to reduce headlight glare at night to windows that allow controlled amounts of light into a room.

Recently, pigments have been discovered that can change color as the temperature changes. They are called thermochromic pigments. They are very new, but applications for them are already being developed. For example, they may be used to help save energy by being applied to roofs. As the pigments become lighter in strong sunlight, they will reflect more heat and help to keep the building cool. But on cooler, overcast days when there is less sunlight, the pigments will become darker and absorb more heat. Thermochromic pigments are used in the clothing industry as well. You may already have seen someone wearing a piece of thermochromic clothing such as a T-shirt.

Many animals are affected by the period of daylight too. Migratory birds do not depart for warmer climates until the hours of daylight get short enough. Mammals that hibernate, such as bears, also respond to shorter day lengths. Courtship in birds is also often stimulated by sufficient hours of daylight, since the female bird wants to lay her eggs and rear her young in the spring and summer when there is abundant food. The way that animals detect light is not fully understood, but it may involve a part of the brain called the pineal gland.

In spring the feathers of the male mandarin duck are very bright and colorful. The feathers are displayed to the female.

The pineal gland is present in mammals, where it produces hormones (chemical messengers) that travel around the body and cause a particular effect. One of these hormones is melatonin, which is released on exposure to darkness. Melatonin affects processes involved with sleep, the onset of puberty and breeding cycles.

A reptile, the primitive Tuatara lizard from New Zealand, is the only animal in existence today that has three eyes. It has a third eye, called a pineal eye, situated on the top of its head. Although covered by skin, the pineal eye has a lens, a retina and a nerve that connects the eye to the brain. It is thought that this third eye is light-sensitive and that it controls certain aspects of the lizard's behavior.

USING LIGHT

EXPERIMENT

Floral clocks

It is possible to tell the time of day by looking at flowers. Some flowers will even open and close their petals at a particular time during the day. Insects and bats soon learn when a flower will be open, and time their visits accordingly. The famous 18th-century biologist, Carl Linnaeus, showed how these flowers could be used as a floral clocks. Because the actual times of dawn and dusk vary in different parts of the country, and flowers are sensitive to other factors such as shading, floral clocks are only accurate in a small area such as a garden or park.

This experiment can only be carried out in summer when there are a large number of plants in bloom. You will need to be up early to discover which flowers open in the early morning. Remember that some flowers only open in the evening. Study the flowers all day so that you know when the different species close their flowers. Record the time at which a flower opens or closes. Go back over the next few days to check that the time is the same. Look particularly for these flowers – dandelion, passion flower, carnation, scarlet pimpernel, hawkbit, bindweed, water lily and evening primrose. Some flowers have regular opening times while others have regular closing times!

Once you have established a regular pattern of opening and closing, you could look for other factors that might affect the pattern. What happens on a cold or wet day?

Key words
Phototropism the growth of a plant shoot toward a light source.
Thermochromic changing color in response to temperature changes.

Nowadays it is common for people to travel around the world in airplanes, passing rapidly from one time zone to another. This can cause a person to suffer from a condition called jet lag. This happens when our body clock is adjusted to the time zone that we have left rather than the one we have arrived in. If the time difference is large, the body tries to behave as though it were asleep just when the person needs to be most active, and vice versa. It can take several days to adjust fully to a new time zone. However, new research suggests that jet lag can be reduced by sitting under bright lights. This seems to help the body to adjust more quickly. Another method being investigated is to take tablets containing melatonin, which tricks the body into resetting its internal clock.

DESIGNS IN SCIENCE

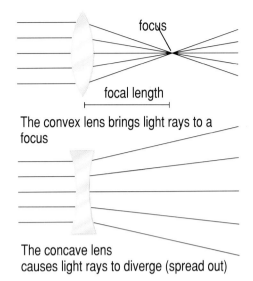

The convex lens brings light rays to a focus

The concave lens causes light rays to diverge (spread out)

Seeing with light

An eye, a camera and a magnifying glass all have one thing in common – the convex lens. Convex lenses are specially shaped pieces of glass or similar transparent material used to form a small, upside-down image of an object. A convex lens is thicker in the middle than at the edges and, because of this shape, it bends light in a special way. When light from an object passes through a convex lens, the rays of light will be bent so that they converge (come together). The point where the light rays converge is called the focus. The focus is the point at which the image will be the most sharp. If something is out of focus, the image will be blurred.

The eye

The eye of vertebrates (fish, amphibians, reptiles, birds and mammals) is probably the most advanced type of animal eye. The eye is spherical in shape and consists of several layers of cells.

Eyes have to be able to focus on objects both near and far away. This is made possible by the ability of the lens to change shape. The light entering the eye must be brought to a focus on the light-sensitive layer, the retina. As light passes from the air through the cornea, the rays are refracted. This achieves most of the focusing. But the fine adjustments to the focus are made by changing the shape of the lens in the eye. Muscles cause the lens to become fatter and more convex or thinner and less convex. The more convex the lens, the more it will bend light. Light rays that come from distant objects are traveling almost parallel to one another when they reach the eye. They require little bending to come to a focus on the retina, so the lens in the eye will be long and thin. However, light rays coming from a close object, such as this book, will be diverging (spreading out) as they reach the eye.

Near-sighted people have blurred distance vision because the image falls in front of the retina. What type of lens would be used to correct this?

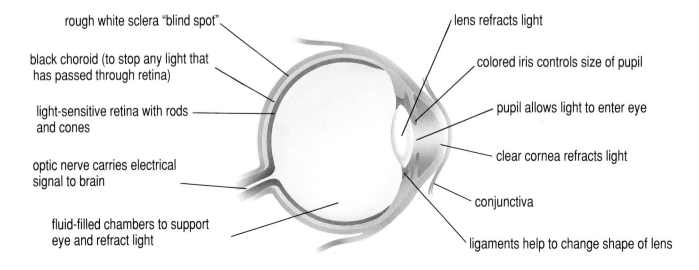

USING LIGHT | 33

These rays will have to be bent quite a lot to come to a focus and a more convex lens will be needed. These changes in the shape of the lens are called accommodation. It happens automatically and very rapidly.

Humans can focus on objects at a wider range of distances than can many other mammals. Dogs, for example, can focus on only a narrow range of distances. Children can focus on a wider range than adults, for the range decreases as we get older. Other animals have developed a variety of ways of focusing on a greater range of distances. Some birds, such as pigeons, can bend their cornea to increase refraction, and so help to bring an object into focus. The squid can actually move its lens toward or away from the retina. Diving birds, such as the cormorant, have extra muscles around their lens that squeeze it to give extra focusing range.

Focused light falls on the retina, a light-sensitive layer that contains two types of cells known as rods and cones. The rods are more numerous and enable us to see in black and white and in dim light. The cones are responsible for color vision. There are three types of cones, one sensitive to red, another to blue and the third to green – the three primary colors of light. The combined action of these cones enables us to see the colors of the visible spectrum (see page 7). However, the cones only work in bright light. At night we rely on rods for our sight and can only see in black and white.

Light falling on the retina excites the rods and cones, which send electrical signals along the optic nerve to the brain. The image is formed upside-down on the retina.

When the brain receives all the signals, the image is put together and turned the right way up. There is one tiny area on the retina that does not have any light-sensitive cells. This is called the "blind spot" and it is where the optic nerve leaves the back of the eye.

The image produced by one eye is very slightly different to the image seen by the other eye, because each is looking from a different position. The brain compares both images and merges them, enabling us to estimate distances. The brain produces a three-dimensional picture. You can discover this by covering one eye and trying to judge distances with the other.

Animals that are preyed upon, such as this rabbit (top), have eyes at the side of their head to give them good all-around vision. The eyes of the owl (right), a predator, are at the front of the head. This gives the owl some 3-D vision, enabling it to catch small animals.

! *Humans have 200,000 light receptors per mm^2, but the buzzard has more than 1 million per mm^2.*

! *The eyes of the net-casting spider are 19 times more sensitive to light than the human eye.*

34 DESIGNS IN SCIENCE

EXPERIMENT

The pupil reflex

It is important to control the amount of light entering the eye since too much light could damage the retina. It is the pupil, a tiny hole in front of the iris, that is responsible for controlling the amount of light entering the eye. If the pupil is large more light can enter, but if the pupil contracts (gets smaller), less light can enter.

In this experiment you will see how your pupil changes size to control how much light enters the eye. You will need a small flat mirror about 10 cm x 10 cm (4"x 4") — such as a make-up mirror — and a small flashlight.

1 Position the mirror so that you can see an image of your eye in the mirror.
2 Hold the flashlight in your other hand and position it just behind and to the side of the mirror, so that the light shines around the edge of the mirror into your eye. Do not shine the light directly into your eye!
3 Watch for any changes in the size of your pupil. What happens to it when you take the light away?

Is your pupil larger or smaller in bright light? Why is this so?

Color vision

Frogs' eyes are sensitive to blue. In times of danger they automatically jump toward the nearest blue area, usually a pool of water!

Humans have cells in their retina that are sensitive to three colors of light: red, green and blue. But not all mammals have color vision. Mammals such as dogs and cats only have very limited color vision. Birds have the best developed color vision of any animal. They have five types of cones, each sensitive to a different wavelength of light. Each cone also contains a colored droplet of oil. The droplet acts as a filter, blocking out some wavelengths of light. This means that the cone can react to a very narrow range of wavelengths of light and this enhances its sensitivity. This enables birds to distinguish between a huge range of colors and shades. However, the pigments and filters are also adapted to the needs of an individual species of bird. For example, sea birds use red filters to cut out the scattered blue light from the surface of the water. Birds such as swallows and swifts that catch their prey, insects, on the wing have lost their red sensitivity. Red filters darken the sky, which would make hunting difficult.

Sight is only associated with animals, but recently scientists have discovered sensitivity to light in microscopic algae. They found that the single-celled alga called chlamydomonas swims

USING LIGHT 35

Some people are red-green color blind. This means they cannot distinguish between red and green colors. How might this affect their everyday life?

toward a light source, guided by a pair of beating hairs. Each hair has a tiny light-sensitive spot. The light is detected by a chemical called rhodopsin. This chemical is also found in the rods of the retina of the vertebrate eye.

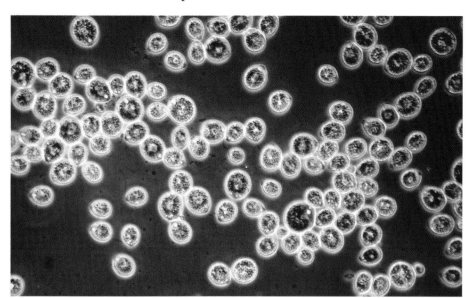

Scientists believe that even simple plants, such as this alga, are sensitive to light.

Cameras

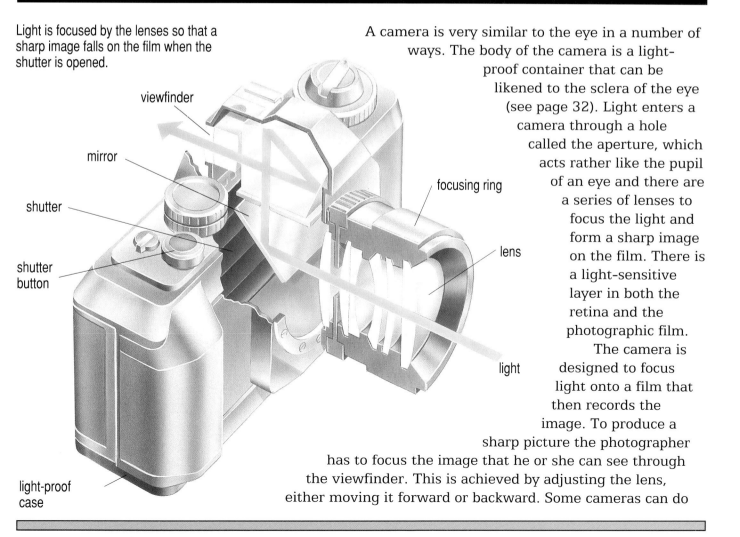

Light is focused by the lenses so that a sharp image falls on the film when the shutter is opened.

A camera is very similar to the eye in a number of ways. The body of the camera is a light-proof container that can be likened to the sclera of the eye (see page 32). Light enters a camera through a hole called the aperture, which acts rather like the pupil of an eye and there are a series of lenses to focus the light and form a sharp image on the film. There is a light-sensitive layer in both the retina and the photographic film. The camera is designed to focus light onto a film that then records the image. To produce a sharp picture the photographer has to focus the image that he or she can see through the viewfinder. This is achieved by adjusting the lens, either moving it forward or backward. Some cameras can do

36 DESIGNS IN SCIENCE

this automatically. The light does not fall on the film at this point. Instead the light passes through the lens and is reflected from a mirror onto the viewfinder. To take the picture, the photographer presses on the button that opens the shutter. The open shutter lets a small amount of light into the camera and then closes again. As the button is pressed, the mirror is flipped out of the way, allowing the light to fall on the film. The time during which the film is exposed to the light is called the shutter speed. Some cameras allow the shutter speed to be adjusted, so that the right amount of light falls onto the film to get the correct exposure. Shutter speeds in daylight are often about 1/125th of a second, but can be as short as 1/2000th of a second in very bright light or as long as 4 seconds when taking photographs at night.

When light falls on the film it produces a change in a light-sensitive chemical coating. These changes are fixed when the film is developed (see page 17).

Television cameras work in a slightly different way than photographic cameras. The thousands of light-sensitive detectors of a television camera are arranged in a grid. Each detector is called a picture cell, or pixel. As light falls on each pixel, it produces an electrical signal. Electronics in the camera scan each pixel many times a second, and the signal from each pixel is sent down a cable in the form of a long coded message. The messages from all the pixels are combined to form a complete picture. The

There are approximately 500,000 pixels on a television screen.

Concave and convex lenses are used in many everyday objects. How many different uses can you think of? What type of lens is used in each case?

The television screen is covered in rows of tiny dots in blue, green and red. Each dot is called a pixel.

The TV camera also uses pixels. Light falling on the pixels produces electrical signals that can be broadcast around the world.

USING LIGHT

EXPERIMENT

A pinhole "camera"

In this experiment you will make a very simple "camera," although you will not use photographic film to record a picture. You will need a cardboard tube no more than 10 cm (4") in diameter and 10 cm (4") long, a pin, some wax paper, some metal foil, rubber bands and a pair of scissors.

1 Cover one end of the tube with the metal foil and the other with wax paper, making sure both materials are stretched tightly over the ends. Secure them in place with rubber bands. The wax paper will be your screen.
2 Make a tiny hole in the foil with the pin. Make sure it is as small as possible.
3 Turn off any lights in the room and hold the "camera" with the foil end pointing toward the window. Place an object on the window sill. Look into the back of the "camera." What can you see? Which way up is the image?
4 Repeat this experiment again, but this time make the hole in the foil much larger. How does this affect the quality of the image?
How could you improve the design of this "camera"?

You will see the image where the spot of light apears in this picture.

The eye of the nautilus, a mollusk, is like a pinhole camera. There is no lens, a retina and an opening.

signal is either recorded or sent out as a broadcast to be picked up by a television set. The set then reconstructs the picture using the same grid of pixels and displays it on a television screen.

The television works by mixing the three primary colors of light: red, blue and green. These are the same as the color pigments in the cones of the eye. A television screen is covered in rows of tiny pixels. Each tiny pixel can be made to glow. There is a row of green pixels, then one of red pixels and so on. The television then fools the eye into seeing colors that are not really there. If neighboring blue, red and green pixels glow, the eye will see white. If a red and green pixel glow, a yellow color will be seen (see page 14). The brightness of each pixel can be altered too, giving a greater range of colors and shades. Televisions in the United States have 525 lines of pixels and these are scanned so quickly that 30 complete pictures are transmitted every second. This enables a continuous moving picture to be built up.

Some animals, such as the crustacean copilia, have evolved eyes that work on a principle similar to a television camera. Copilia has a lens that forms an image on the retina. The retina only has 9 receptors present, so the image is of very poor quality. A second lens and retina then scans the image at least 10 times, allowing a better picture to be built up.

Key words
Convex curved out; having a bulging shape.
Focus a point at which light rays meet.
Lens a transparent substance with curved surfaces for bending light rays.

38 DESIGNS IN SCIENCE

Invisible light

The goldfish is able to see both UV and IR wavelengths of light.

Not all wavelengths of light are visible; some are invisible. The reason has to do with the amount of energy in the different wavelengths of light. The shorter the wavelength, the greater the amount of energy carried by the photons (see page 7). For example, blue photons carry more energy than red photons. Infrared (IR) photons do not carry much energy, far less than even red light photons. They do not have enough energy to excite the cones in the human retina, so infrared light is invisible to our eyes. Ultraviolet light (UV) is also invisible to our eyes, but this is because ultraviolet photons carry too much energy and this could damage our retina. The pupil of the eye and our eyelids both help to shield our eyes from the UV light present in sunlight.

Ultraviolet light

Sunlight contains significant amounts of UV light. It is important to be aware of how UV light can affect people and other animals and of how it can be used.

UV light can be reflected strongly from the surface of water and from white surfaces such as snow. Snow blindness is caused by the eyes being exposed to bright light that contains ultraviolet. It can affect skiers and people living in polar regions, such as the Inuits. They need to wear protective goggles or sunglasses that filter out the UV light to prevent serious eye damage. The increasing damage done to the ozone layer in the earth's upper atmosphere means that more and more UV light is now reaching the surface of the earth (see page 14). As a result, skin cancers and eye problems are becoming more common. In some parts of the world, within the next decade or so, it may be necessary to wear glasses all the time in order to cut out the invisible UV light.

Climbers have to wear protective goggles to shield their eyes from the UV light that is reflected from the snow.

UV light emitted by some artificial lights as well as in sunlight can cause oil paintings to darken and watercolors to fade by breaking down the pigments in the paint. It can even cause the threads of silk to weaken and break.

However, UV light can also be very useful. For example, it can be used to sterilize objects or in photochemical processes (see page 27). U.S. scientists are experimenting with solar energy, using the UV wavelengths in sunlight to treat industrial effluent (liquid waste). The effluent is pumped along glass tubes that are exposed to UV

USING LIGHT 39

A flower such as this potentilla (top) looks very different when seen under UV light (right).

light. By the time the water gets to the end of the tube, all the pollutants in the water have been broken down by the UV light into harmless compounds.

The human eye cannot see UV light, but some insects, such as the honeybee, can do so. However, bees cannot see some of the colors that we can see, such as red. Their eyes are most sensitive in the blue-green and ultraviolet parts of the spectrum. Bees have been found to discriminate between yellow, blue-green, blue, violet, UV and "bees purple" (a mix between yellow and UV). In fact, they see a very different color spectrum than that seen by humans.

Bees and other insects are very important to plants because they transfer pollen as they fly from one flower to another. This pollinates the flower and so allows fertilization and seed production to take place. However, plants first have to attract the bees to their flowers. Plants do this by having flowers with large colorful petals as well as producing scent and nectar. Many flowers are bright red, but bees cannot see red. So red flowers also reflect UV light that the bees can see. Some flowers have evolved eye-catching patterns of spots or lines called honey guides that are designed to lead the bee to the sugary nectar and protein-rich pollen within the flower.

Many minerals, foods, common chemicals and even living organisms emit fluorescent light when they are lit by UV light. The effect of fluorescence is used by scientists to make washed clothes appear whiter. Chemicals that fluoresce under UV light, such as calcofluor, are used as optical brighteners in laundry detergents. They bind to cellulose fibers, so clothes made of natural fibers like cotton have a blue-white appearance in sunlight.

Some types of coral can fluoresce when they are hit by UV light.

40 DESIGNS IN SCIENCE

EXPERIMENT

The honeybee's favorite color

In this experiment you will train bees to visit containers of sugar solution that are placed on a particular color of card. You will need a number of cards of different color paper such as white, black, red, blue and several different shades of gray. You will also need some small dishes, each containing a little sugar solution.*

1 On a sunny day in summer when the bees are active, go outside and place your blue card in a sunny position. Place a dish of sugar solution on it. Leave it there for a day, so the bees learn that sugar is to be found on the blue card.

2 The next day, set out your blue card together with the white, black and gray cards arranged randomly nearby. Do not place sugar solution on any of the cards.

3 Wait and see which card the bees fly to. They should be able to distinguish between the different colors and will fly to the blue card first, hoping to find food.

4 Repeat this experiment, but this time place sugar on all the cards. Which color do the bees prefer? Watch the bees while they are visiting flowers. Which color of flower has the most bee visitors?

Be careful not to get too close to the bees or disturb them or they may sting you.

Infrared light

When materials become very hot, the molecules in them vibrate so rapidly that they emit visible light.

Infrared light has a wavelength that is just longer than that of visible red light. Nearly everything in the natural world emits some IR light. So do many human-made objects, especially hot ones such as stoves or ovens. This light comes from molecules that are vibrating rapidly. As the molecules warm up, they gain energy and they begin to vibrate. Some IR light is emitted, but the wavelength lies outside the visible spectrum, so humans cannot see it. If even more energy is added, the molecules vibrate more and more rapidly, causing the wavelength of the light emitted to become shorter. Eventually, the object becomes so hot that it also starts to emit light of wavelengths short enough to be seen by the human eye. The object is then described as being "red hot."

Many animals have the ability to detect IR wavelengths of light. For example, the piranha fish has a fearsome reputation as a very efficient hunter. Its skill as a hunter is largely due to its ability to detect longer wavelengths of light than most other fish. It lives in the murky waters of the Amazon river in South America. These waters are almost impenetrable to visible light.

USING LIGHT

However, IR wavelengths do manage to penetrate the gloom, and the piranha has evolved eyes that allow it to detect the IR given off by its prey.

Similar IR-sensitive systems are found in military weapons. IR searchlights and IR-sensitive goggles give a soldier a better view at night. Military aircraft such as the F-117 Stealth Fighter and the Apache helicopter use IR light to see in the dark. One IR-detecting system is known as Forward Looking Infra-Red (FLIR). It consists of a large lens that looks like a giant eyeball that collects and focuses IR light onto sensors. The sensors then produce a display on a screen for the pilot or observer. Military sensors are expensive, but they are now being adapted to help civilian aircraft pilots carry out difficult tasks. Sensors are useful, for example, when an aircraft is taxiing on the ground in bad weather conditions. The plane's navigation systems are not designed to work on the ground, but IR systems that can "see through" rain and fog can help pilots see where to go when visibility is poor.

The piranha fish can see infrared light. It can hunt its prey in dark waters.

! *Camouflage-detecting cameras using IR can discriminate a less than 1°C (1.8°F) difference in temperature between an object and its background to show the presence of a human being.*

The IR systems described above are only recent developments, but an IR night sight has been used by a deep sea fish for hundreds of thousands of years. The fish, pachystomias, lives in the deep, dark waters of the oceans where only blue wavelengths of light can penetrate. Most fish at these depths have eyes that are sensitive only to blue. Pachystomias, however, can produce a red beam of light, which it uses to light up the water as it swims around looking for its prey. None of its prey can see the red light, but pachystomias can, so it has a very effective hunting weapon.

Scientists are also developing microscopic antennae, the size of a grain of sand, that are capable of detecting IR light from missiles or from emissions of global warming gases, such as carbon dioxide and methane. The designs are based on the antennae that are used by some insects to detect IR and "see in the dark."

The FLIR system is installed in police helicopters to help find people in darkness or undergrowth.

Key words
Fluorescence the emission of light by a material when hit by ultraviolet light.
Infrared invisible wavelengths of light lying beyond the red end of the visible spectrum.
Ultraviolet invisible wavelengths of light, lying beyond the blue end of the visible spectrum.

DESIGNS IN SCIENCE

Interference and polarized light

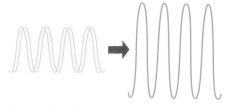

Positive interference creates more intense color.

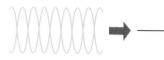

Negative interference cancels out the peaks and troughs and creates darkness.

Light waves spread out in ever-increasing circles, just as ripples spread over the surface of a pond into which a pebble has been tossed. If two pebbles are thrown in, the spreading ripples from each will meet. When the crests of two waves meet, they combine to form an even bigger crest, and where two troughs meet they make a bigger trough. But where a crest and a trough meet, they cancel each other out. Light waves behave in exactly the same way.

Interference patterns

Interference patterns are used by a number of insects to create their characteristic colors. Butterflies, for example, have an iridescent (rainbow-like) sheen to their wings that changes slightly with the angle of view. The colored patterns are produced by tiny scales, many of which are in fact colorless, that cover the wings. When viewed under a microscope, it is possible to see that each scale has tiny ridges and furrows in its surface that cause the light to be scattered (see page 9) as it falls on the scales. As light is reflected from the scales, different colors are produced as the waves of light interfere with one another.

It is very easy to see interference patterns of light by looking at the surface of a compact disc. The surface of a compact disc is covered by tiny pits that scatter light, just like the scales of a butterfly's wing. When the disc is tilted to the light (either sunlight or artificial light), iridescent bands of color can be seen. These bands of color alter as the angle of the disc is changed. Interference also causes the rainbow-colored bands that can be seen on soap bubbles. These bands are produced when light is reflected from both the outer and inner surface of the bubble. Light reflected from the inside of the bubble has to travel slightly further to your eye than light reflected from the outside. So when light from the inside joins with light reflected from the outside, the light waves are out of step and the crests and troughs of the waves interfere positively.

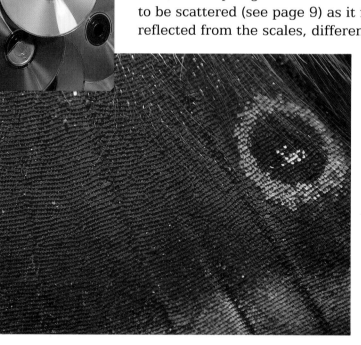

The pits on the surface of a compact disc (above) and the tiny scales on a butterfly's wing (right) both scatter light and produce iridescent patterns.

USING LIGHT 43

Polarized light

The waterboatman, a type of insect, lives in ponds. Its eyes do not cut out polarized light but actually enhance it, so that it can spot light reflecting from the surface of water.

Why do photographers use polarizing filters on their cameras?

Light from a source such as a light bulb is composed of waves of light that vibrate in many different directions. Polarized light, on the other hand, is light that consists of waves vibrating in one direction only. To produce polarized light, the light from a normal source has to be passed through a special filter that only lets through light vibrating in one direction.

Polarizing sunglasses are designed to cut down the reflected glare from shiny surfaces, such as water or glass. The reflected light is polarized, which means that it is vibrating in one direction. Polarizing sunglasses are made from polarized film that is vibrating in another direction, so they cut out most of the reflected light. This reduces the glare from the shiny surfaces, but also lowers the overall light level.

You can easily try this for yourself. All you need is a pair of polarizing sunglasses. Look at a river or pond in bright light without the glasses. You will probably be unable to see much beneath the surface. If you put on the sunglasses, you will see that the colors become more intense and that you can now see below the surface of the water. However, if you rotate the sunglasses through 90° so that they are vertical and look again through one of the lenses, you will see that the glare still remains. This is because the film and the glare are now vibrating in the same direction and so the glasses let the light through. The polarizing filter only works at the correct angle.

Even sunlight from a clear blue sky is slightly polarized by the atmosphere. Honeybees are able to determine the sun's position even if it is obscured by clouds, by studying the polarized light in a patch of blue sky. This ability to navigate using polarized light is called astrotaxis. The bee can see a sky map with patterns created as the atmosphere polarizes the sun's light. The bee aligns itself at a particular angle to the sun, compensating automatically for the movements of the sun. As a result the bee can maintain a constant compass direction throughout the day and can find its way back to the hive.

Bees are not the only insects to be able to detect polarized light. Some other insects, such as fruit flies and some beetles, have similar abilities. Interference of polarized light is particularly important in engineering. It is possible to reveal stresses in transparent materials such as plastic by using polarized light. The polarized light is shone through a plastic object. The light is then reflected from contours and deformities within the plastic. Interference patterns are seen as bands of colored light. The more bands there are, the more the material is under stress.

Polarized light is used to show up stress points in materials such as plastic. The stress points appear as bands of color.

Key words
Interference the meeting or interaction of two or more light waves.
Laser a form of light that contains waves of the same wavelength, with the crests and troughs aligned. Laser stands for Light Amplification by Stimulated Emission of Radiation.
Polarized light wavelengths of light that vibrate in one direction only.

The future

It is always difficult to predict the future. However, we can expect to see many novel developments in the use of light over the next decade or so. Efforts to produce cheaper and more efficient forms of lighting will certainly continue. Further interesting research is taking place into the use of laser light for computing and communications, and into the manufacture of materials that can alter their color under different physical conditions.

Electric lighting is becoming progressively more efficient, and much less energy is being wasted in the form of heat. Compact fluorescent bulbs are already available in the shops. The latest kinds of compact, energy-efficient fluorescent bulbs even look like conventional tungsten ones, although they work on a different principle.

Compact fluorescent bulbs are much more efficient than the tungsten bulbs we currently use. A 100 watt tungsten bulb can be replaced by a 25 watt fluorescent; 75 percent of the electricity is saved.

The next generation of bulbs, still under development, does not even have a filament or electrode. Nor is the mercury vapor within the bulb excited by an electric current, as it is in fluorescent lighting. Instead, a special coil in the center of the bulb emits high frequency radio waves. These radio waves excite the vapor within the bulb, which then emits UV light. This UV light, in turn, excites the phosphor coating inside the bulb. The light bulb dims as the phosphor ages, but these bulbs will last for up to 60,000 hours. That is about 14 years of normal use! They do not require a starter and will fit into existing sockets.

Experiments are being carried out by Russian spacecraft to turn night into day by reflecting the sun's rays toward earth. Huge solar mirrors may one day be unfurled in space. An experiment using a smaller prototype was recently carried out. The latest plans involve 100 solar reflectors positioned in a ring around the earth, at altitudes ranging from 1,550 to 5,530 km (969 to 3,456 mi.). The reflected light could illuminate cities or construction sites in the north of Russia during the long Arctic winter, when the sun never rises above the horizon. The space mirrors could also be used to provide light for disaster areas to assist rescue teams at night.

Lasers are quite familiar to us nowadays, in the form of thin beams of light that can travel for long distances without spreading out. Light from an ordinary source, such as a light bulb, consists of a jumble of different wavelengths. Laser light, on the other hand, is a very pure and ordered form of light. A laser is created by making light waves oscillate rapidly back and forth. The waves of light are all the same length and all the troughs and crests of the waves are aligned. Light ordered in this way is said to be coherent. Powerful lasers can concentrate a fine beam of

USING LIGHT 45

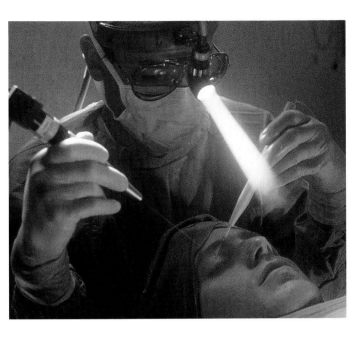

Laser beams can be finely controlled, so they are suitable for use in surgery. They are often used in delicate eye operations or to cut through blocked arteries.

intense light energy into one spot. This light energy can be used to cut through metals and other materials.

Most modern cameras still work with light-sensitive film, but in the future we may be using digital photography. In this process, the image will no longer be stored on film but on a CD, which is read by laser light. There will be no need to use light-sensitive chemicals, either at the stage of capturing the image or when developing it. The camera will be similar to a conventional camera, but when the shutter opens the image will be recorded as an array of colored dots on a special disc. The data will be transferred to the CD and stored for future use.

Biologists continue to discover more and more about the effect of light on plants. Flower growers benefit from this knowledge by controlling the lighting in greenhouses so that plants produce flowers at exactly the right time for their market. Soon growers will be able to control the growth pattern of plants by altering the periods of light and dark together with the temperature. Plant breeding currently takes a long time, since it is limited by the time it takes for a plant to grow, flower and produce seed. However, by using artificial environments where the length and quality of light can be controlled as well as temperature and nutrient supply, it is possible to reduce this time considerably. For example, it is now possible to grow up to four generations of wheat in one year. Some of the rapid-cycling brassicas (types of cabbage, broccoli and turnip), complete their whole life cycle in just 36 days when grown under 24 hours of fluorescent light. This research means that new varieties of crops can be produced more quickly, which will benefit the whole of humankind. For example, we will see new varieties of wheat that require less fertilizer for the same yield, that are more resistant to disease or that can grow in a greater range of climates.

Light affects animals too, in ways that we are only just beginning to understand. When we learn more about how light affects us, we will be able to use it more effectively. Even the color of the light may be important. The ancient Egyptians studied the effects of color on people. They painted rooms in particular colors in order to treat certain ailments. The science of chromotherapy (using particular wavelengths of light to treat disease) is rediscovering some of these lost secrets. There are many more fascinating ways in which light can be used just waiting to be discovered. Many of these will undoubtedly be found in the natural world.

The conditions in a growth chamber can be carefully controlled to give the ideal light levels for plant growth.

Glossary

atom the smallest particle of a chemical element that retains that element's properties.
bioluminescence the giving off of light by living organisms.
catalyst a chemical that increases the rate of a reaction without being changed itself.
chlorophyll the green pigment found in plants that enables them to photosynthesize.
concave curving inward: having a dish shape.
convex curving outward: having a bowl or bulging shape.
electron a tiny negatively charged particle of an atom.
emulsion a milky liquid with oily particles in suspension.
filter a transparent substance that will only let through light of one or more particular wavelengths.
fluorescence the emission of light by a material when hit by UV light.
focus a point at which light rays meet.
image a reflection formed by a mirror or lens, an exact copy of an object.
infrared the invisible wavelengths of light lying beyond the red end of the visible spectrum.
interference the meeting or interaction of two or more light waves.
laser a form of light that contains waves of the same wavelength, with their peaks and troughs aligned.
lens a transparent substance with curved surfaces used for bending light rays.
mirror a smooth polished surface that reflects light in a regular way to form an image.
molecule the smallest naturally occurring particle of a chemical compound made from two or more atoms.
neutron a particle with no electrical charge found in the nucleus of an atom.
opaque not allowing light to pass through.
photochemistry a process that uses light to cause a chemical reaction.
photon a tiny unit of light that has a small amount of energy.
photosynthesis the process by which plants make food using light energy.
phototropism the growth of a plant shoot toward a light source.
pigment a substance that gives a material a particular color.
polarized light wavelengths of light that vibrate in one direction only.
reflection a wave of light that bounces off an object.
refraction the bending of light rays as they pass from one medium to another.
spectrum the rainbow colors produced when light passes through a raindrop or a prism (specially shaped piece of glass).
thermochromic changing color in response to temperature changes.
ultraviolet the invisible wavelengths of light lying beyond the blue end of the visible spectrum.
wavelength the distance between two neighboring peaks or troughs of a wave.

Answers to the questions

p. 9   Translucent — sunglasses, some types of glass such as frosted glass.
p.10   Concave — shaving and make-up mirror; convex — driving mirrors.
p. 14   To protect skin from damaging UV rays in sunlight.
p. 19   B/w film is not sensitive to red light.
p. 21   Lights used depends on the home.
p. 21   They can be used in most fittings such as lamps and reading lights, ceiling lights, etc. They use less energy, last longer, require fewer raw materials.
p. 23   By using chemicals called pheromones that the insect can detect.
p. 29   Frogs, squid, cuttlefish, octopus.
p. 30   To absorb and reflect heat, to show hot spots or overheating, to show excessive cooling, to show the emotional state of the wearer, thermometers.
p. 32   Concave lens to cause light rays to diverge slightly so focus further away from the lens.
p. 35   Red-green colorblindness — inability to distinguish between red and green on traffic lights, maps such as subway and road maps, clothing.
p. 36   Concave and convex both used in glasses — depends on sight defect, convex in camera lens, telescopes, microscopes, magnifying glass.
p. 43   To enhance colors and eliminate glare from reflective surfaces such as water.

Index

Key words appear in **boldface** type.

A
absorption 14, 16, 19, 24, 30
algae 34–35
amphibians 32
aniline 16
animals *See also specific groups of animals (e.g., birds)*
  behavioral patterns 28, 30, 45
  energy sources 24
  pigments used by 16
  vision 32–35, 37
antennas 41
Arctic region 29
argon 20–21
artificial light 6–7, 20–23, 38, 44–45
astronomers 10
astrotaxis 43
atmosphere 24–25, 43
atoms 6, 17, 20, 23, 27, 46

B
bacteria 23
bats 31
bees 39, 43
beetles 23, 43
bending light *See* refraction
bioluminescence 22–23, 46
birds 15–16, 30, 32–34
black (color) 18–19, 29
black-and-white film 18–19
black-and-white vision 33
"blind spot" 33
blue (color)
  in artificial light 22
  and bee vision 39
  and human vision 18, 33–34
  in ocean depths 11, 41
  in photographic film 18
  pigments 16, 29–30
  as primary color 14, 33
  and sea bird vision 15, 34
  in television screens 37
  as tree frog true color 14–15
  in visible spectrum 7
body clock 31
brain 13, 16, 29–30, 32–33
breeding cycles 30
brown (color) 16
bulbs
  fluorescent 44
  tungsten filament/tungsten halogen 20–21, 44

C
calculators 26
cameras 11, 28, 35–37, 41, 43, 45
camouflage 16, 29, 41
cancer *See* skin cancer
carbon dioxide 24–27, 41
catalysts 27, 46
CDs *See* compact discs
cells
  light-producing 23
  light-sensitive 28, 32–35, 37
  photosynthesis in 24–26
  pigments in 16, 29
  respiration in 24
chameleons 29
chemical reactions *See* bioluminescence; photochemistry; photosynthesis
chlorophyll 24, 46
chloroplasts 24
chromotherapy 45
clothing 30, 39
color film 17–18
color printing 18
colors 8, 16, 45–46 *See also* filters; pigments; *specific colors (e.g.,* red*)*
color vision 33–35
color wheel 14
communication 16–17, 22–23
compact discs (CDs) 42, 45
computers 44
concave
  definition of 46
  lenses 36
  mirrors 10
cones (eye) *See* rods and cones
convex
  definition of 37, 46
  lenses 32–33, 36
  mirrors 10
cornea 32–33
cosmic rays 8
courtship 16, 22, 30
cyan (color) 14, 18–19

D
day length 28, 30
digital photography 45
dyes 16, 18–19

E
electricity
  generation of (using solar panels) 26
  transmitted by nerves 13, 32–33
electric lights 20–22, 44–45
electrochromic glass 30
electrolysis 27
electromagnetic spectrum 6–8 *See also* visible spectrum
electronic scanners 19
electrons 8, 20, 46
emulsions 17–18, 46
endoscopes 13
energy 6, 20, 38 *See also* electricity; heat; solar energy
energy efficiency 21, 44
engineers 13
experiments
  color mixing 15
  floral clocks 31
  honeybees 39
  periscopes 10
  photosynthesis 25
  pinhole camera 37
  pupil reflex 34
  total internal reflection 13
exposure (photography) 28, 36
eye *See also* vision
  camera compared to 35
  photographic film compared to human 17
  pineal 30
  pupil reflex 34
  structure and function 32–33
  ultraviolet damage to 14, 38
eyeglasses 11–12 *See also* sunglasses
eyelids 38

F
feathers 16, 30
fiber optics 12–13
filaments 20–21
film (photography) 17–19, 35–36, 45
filters 14–15
  in bird eyes 15, 34
  definition of 19, 46
  in fish scales 11
  in photography 18–19
  polarizing 43
fireflies 6, 22–23
fish 10–12, 23, 32, 38, 40–41
fixer (photography) 17
flowers 28–29, 31, 39, 45
fluorescence 39, 41, 46
fluorescent lights
  bulbs 44
  plants grown under 45
  tubes 20–22
focus 32, 35, 37, 46
food chain 24
Forward Looking Infra-Red (FLIR) system 41
frogs 14–16, 34

G
gamma rays 7–8
gases 6, 20, 41 *See also specific gases (e.g.,* oxygen*)*
glass
  in bulbs 20–21
  in lenses 32
  in mirrors 10
  in optical fibers 12–13
  photochromic 14
  in photographic screens 19
  as reflective medium 9, 43
  thermochromic 29–30
glasses *See* eyeglasses; sunglasses
glowworms 6, 22
green (color)
  and bee vision 39
  and human vision 17, 33–34
  in leaves 24
  in photographic film 18–19
  as primary color 14, 18, 33
  produced by click beetle 23
  in television screens 37
  as tree frog apparent color 14–15
  in visible spectrum 7
greenhouses 45
growth chambers 22, 45

H
halogens 21
hatchet fish 10–11
heat 6, 21, 24, 26–27, 30, 40
hibernation 30
honeybees 39, 43
hormones 30
humans
  behavioral patterns 28, 31
  respiration 25
  vision 33–34, 38–39
hydrogen 27

I
image, definition of 46
indigo (color) 7
infrared (IR) light 6–7, 19, 22, 40–41, 46
insects
  bioluminescence 22–23
  flowers visited by 28, 31
  pigments extracted from 16
  vision 39, 41, 43
interference 42–43, 46
internal clock 31
iridescence 42
iris (eye) 32, 34
IR light *See* infrared light

J
jet lag 31

L

lasers 43–46
leaves 24–26, 28
lenses
  in cameras 11, 35–36
  definition of 37, 46
  in eyes 12, 30, 32, 37
  in weapons systems 41
lighting *See* electric lights
light meters 28
light-producing organs
  in fish 10–11
  in insects 22–23
light sensors 28–31, 41
  in eyes 32–33
light waves 6–8, 42–43
luciferin 22

M

magenta (color) 14, 18–19
magnifying glass 32
mammals 11, 30, 32–34
medicine 13, 45
melanin 14
melatonin 30–31
mercury vapor 21–22
methane 41
microscopes 11
microwaves 6, 8
mirrors 9–11
  adjustable (in cars) 30
  in cameras 35
  definition of 13, 46
  in solar power plants 26
  space (for capturing sunlight) 44
molecules 17–19, 24, 46
muscles 32–33

N

natural pigments 16
negatives (photography) 17–19
neon lights 21
nerves 13, 23, 30
neutrons 20, 46
Newton, Sir Isaac 6
night vision 11, 33
nitrogen 20
nucleus (atom) 20

O

oil droplets 15, 34
opaque materials 9, 46
optical brighteners 39
optical fibers 12–13
opticians 12
optic nerve 32–33
orange (color) 7, 23
oxidation 20–21
oxygen 20, 22–25, 27
ozone layer 14

P

paint 16, 38
periscopes 10
petals 28–29, 39
petioles 26
phosphor powder 21–22, 44
photochemistry 27, 46
photochromatic glass 14
photographic film *See* film
photography 17–19, 28, 43, 45
photolysis 27
photonasty 28
photons 7–8, 17, 20, 38, 46
photosynthesis 24–28, 46
phototropism 28, 31, 46
photovoltaic cells 26
phytochrome 28
pigments 16
  in bird eyes 34
  in cells 16, 24, 29
  definition of 8, 19, 46
  in paintings 38
  thermochromic 30
pineal gland 30
pin hole camera 37
pink (color) 16, 19
pixels 36–37
plants
  grown under artificial light 22, 45
  light energy captured by *See* photosynthesis
  light sensors in 28–29, 46
  pigments in 16, 24
platinum 27
polarized light 43, 46
pollination 28, 39
predators 11, 23, 33–34, 40–41
primary colors 14, 16, 18, 33, 37
printing 17–19
prisms 8, 46
protons 20
pupil (eye) 32, 34, 38

R

radar 6
radio waves 6, 8, 44
rainbows 7–8
receptors (light) *See* light sensors
red (color)
  and bee vision 39
  and human vision 33–34
  in neon light 21
  in photographic film 18–19
  pigments 16
  as primary color 14, 18, 33
  produced by deep sea fish 41
  and sea bird vision 15
  in television screens 37
  in visible spectrum 7
reflection 9–14, 16, 19, 30, 43, 46
refraction 7, 9, 11–13, 32–33, 46
reptiles 29–30
respiration 24–25
retina 30, 32–35, 37
rhodopsin 35
rods and cones (eye) 32–35, 37

S

scales (butterfly wings) 42
scales (fish) 10–11
secondary colors 14
seeing *See* vision
sensors *See* light sensors
shutter (camera) 35–36
silicon 26
silver 14, 17
silver bromide 17
silver halide 17–19
skin 14–16, 18, 29
skin cancer 14, 38
sleep 30
snow 17, 38
snow blindness 38
soap bubbles 42
sodium carbonate 27
sodium vapor 22
solar energy 26–27, 38
space mirrors 44
spectrum 8, 46 *See also* electromagnetic spectrum; visible spectrum
speed of light 11
stars 10
stress 43
sun 6–7, 24, 29, 44
sunglasses 9, 14, 38, 43
suntan lotions 14
surgery 13, 45
synthetic pigments 16

T

tanning 14
tapetum 11
telecommunications 13, 44
telescopes 10
television 6, 36–37
thermochromic
  definition of 31, 46
  pigments 30
three-dimensional vision 33
titanium oxide 27
total internal reflection 12–13
translucent materials 9
transparent materials 9, 11, 19, 32, 37, 43, 46 *See also* glass
trees 25–26
tubes, fluorescent 20–22
tungsten filament/tungsten halogen bulbs 20–21, 44

U

ultraviolet (UV) light 38–40
  in artificial light 21, 44
  damage caused by 14
  definition of 41, 46
  filtering of 14
  as part of electromagnetic spectrum 7–8
  photographic film sensitive to 17
  water broken down by 27
uric acid 23
UV light *See* ultraviolet light

V

vertebrates 32, 35
vibrations 40, 43
viewfinder (camera) 35–36
violet (color) 7, 39
visible spectrum 6–7, 17–18, 33, 40, 46
vision 32–35
  infrared 41
  night 11, 33
  ultraviolet 39
  underwater 12, 40–41

W

warning colors 16
water
  and bird colors 16
  and outdoor photography 17
  in photosynthesis 24
  reflected glare produced by 43
  as refracting medium 7–9, 11
  UV light reflected by 38
  UV light used to break down 27
wavelength 7–8, 46 *See also* filters; photons; pigments; refraction
white
  mixing colors to make 14
  pigments 16, 29
  of sunlight 7–8
  in television screens 37
white (color)
  in artificial light 20, 22
windows 30

X

X rays 7–8

Y

yellow (color)
  in artificial light 20, 22
  and bee vision 39
  in photography 19
  pigments 16, 18, 29
  produced by firefly 23
  as secondary color 14
  in television screens 37
  as tree frog filter 14
  in visible spectrum 7